DIVING AND SNORKELING GUIDE TO

Grand Cayman Island *including* Little Cayman *and* Cayman Brac

by Carl Roessler
and the editors of Pisces Books

 Pisces Books ● New York

Staff

Publisher	**Herb Taylor**
Project Director	**Cora Taylor**
Series Editor	**Steve Blount**
Editors	**Carol Denby**
	Linda Weinraub
Assistant Editor	**Teresa Bonoan**
Art Director	**Richard Liu**
Artists	**Charlene Sison**
	Alton Cook
	Dan Kouw

Publishers Note: At the time of publication of this book, all the information was determined to be as accurate as possible. However, when you use this guide, new construction may have changed land reference points, weather may have altered reef configurations, and some businesses may no longer be functioning. Your assistance in keeping future editions up-to-date will be greatly appreciated.
  Also, please pay particular attention to the diver rating system in this book. Know your limits!

All photographs are by the author.

Library of Congress Cataloging in Publication Data

Roessler, Carl, 1933-
  Diving and snorkeling guide to the Cayman Islands.

  Bibliography: p.
  1. Scuba diving—Cayman Islands—Guide-books. 2. Skin diving—Cayman Islands—Guide-books. 3. Cayman Islands—Description and travel—Guide-books. I. Title
  GV840.S78R59 1984   797.2'3   84-9553
  ISBN 0-86636-034-4

Printed in Hong Kong

10 9 8 7 6 5 4 3 2

Table of Contents

How to Use This Guide

This guide was developed to familiarize you with the location and terrain of the principal dive sites of the Cayman Islands. Because you will be diving with a local hotel, dive boat, or day service in almost all cases, you will not be required to find the site yourself. When the planned dive location is announced, however, you'll have a good idea what to expect in the way of depth, current, topography, and photo subjects. This general familiarity with the dive sites will serve you well in planning your dives for photographic results and in avoiding decompression.

The live-aboard dive vessel Cayman Diver II *is just one way for divers to experience the calm, clear waters of the Cayman Islands.*

The Rating System for Divers and Dives

Our suggestions of the minimum level of expertise required for any given dive should be taken conservatively keeping in mind the old adage about there being old divers and bold divers but few old bold divers. We consider a *novice* to be someone in decent physical condition who has recently completed a basic certification diving course, or a certified diver who has not been diving recently or who has no experience in similar waters. We consider an *intermediate* to be a certified diver in excellent physical condition, who has been diving actively for at least a year following a basic course and has been diving recently in similar waters. We consider an *advanced* diver to be someone who has completed an advanced certification diving course, has been diving recently in similar waters, and is in excellent physical condition. You will have to decide if you are capable of making any particular dive, depending on your level of training, recency of experience, and physical condition, as well as water conditions at the site. Remember that water conditions can change at any time.

It is important not to over-rate your skills when diving at major dropoffs in the Caymans; these islands are noted for the steepness and the depth of their walls. In addition, most of these dropoffs have relatively deep reef crests which keep you at depths of 45–70 feet (15–21 meters) or more throughout your dive. Even experienced divers are careful to monitor depth, time, air supply, and decompression throughout each excursion.

I've organized the guide by area, the way Caymans diving services naturally offer their dives. I describe first the shallow dive areas of the West End, where an overwhelming majority of dive hotels and services are located. These are also the sites snorkelers and shore divers may reach by swimming out on their own.

The second group of dive locations includes sites along the walls of the West End. Again, proximity to the hotels and services of Seven Mile Beach are the common characteristic, but these sites are both deeper and farther from shore. Those two groups of reefs represent the vast majority of all diving done in the Caymans.

The third group is made up of sites located along the fabulous North Wall. These dives are offered infrequently due to their distance and weather considerations. Similarly, the group of destinations on the South Coast are seldom dived in comparison to the West End.

The East End of Grand Cayman faces directly into prevailing winds and currents, so weather plays a crucial role in allowing you to dive any of its reefs.

To end our selection I describe some of the fine diving of Little Cayman and Cayman Brac.

Along the way I try to offer some insights into the islands' services by type, so you can visualize some of the choices you face when selecting different types of dive services.

1

Overview of the Cayman Islands

Looking at today's modern, sophisticated Cayman Islands, it's hard to believe that most of what you see has exploded into life only in the past few years. But these cosmopolitan and highly active islands have a rather sedate, even sleepy history.

They were first discovered in 1503, when the ubiquitous Christopher Columbus was blown off course during his fourth voyage to the New World. Columbus actually landed at Little Cayman and Cayman Brac, which he named *Las Tortugas* after the many sea turtles sighted there. Turtles could survive aboard ship for up to a year, and they were a coveted source of fresh meat for sailors on long voyages. This made the new islands a favorite stop for sailing ships, but over many years it had a severe impact on the turtle population. Only in the past decade have serious efforts been undertaken to breed and protect these endangered sea-going reptiles. Later maps showed the islands named *Lagartos* (large lizards) and by 1530 *Las Caymanas* (a name derived from the Carib Indian word to describe small marine crocodiles).

In 1586 Sir Francis Drake stopped at the Caymans, on an expedition during which he plundered Santo Domingo. Drake's log records the islands as "not inhabited" and infested with "great serpents called *Caymanas.*"

In 1592 Captain William King sailed from Jamaica and found fresh water, turtles, and wildfowl in abundance. By 1643 the islands were a regular provisioning stop for sailing ships in the area.

The Spanish, fearing that they were losing their grip on the Caribbean, began attacking English and French ships. The English response was the *Privateer*, a de facto English Navy of private ships supporting themselves by piracy.

Oliver Cromwell, trying to drive the Spanish out of the Caribbean, attacked Hisponida with 7,000 men and was repulsed. A second attack succeeded in taking Jamaica, which Cromwell hoped to colonize. There were few volunteers, but it is from those original English colonists that the earliest settlers of Cayman were drawn.

A diamond blenny hides among the tentacles of a large anemone. The anemone subdues prey by stinging them with its waving tentacles. For small fish such as the blenny, which are immune to the toxin, the tentacles are a perfect refuge. ▶

Just offshore of famed Seven Mile Beach, north of George Town along the edge of West Bay, is Soto's Reef. The dark patches of coral reef contrast with the white sand bottom.

By 1713 peace was declared, and the legal justification for piracy disappeared. The pirates were completely wiped out by 1730. Because notorious pirates, including Blackbeard, were based in the remote Caymans, there are the inevitable stories of buried treasure. Who knows, someday the media may flash the headline that a treasure actually has been found. Visitors more likely, will discover that the real treasures of the Caymans are fun, sun, and some of the world's greatest reef walls.

In November 1788, a crucial event occurred that has affected all subsequent developments in these islands. It is known as the "wreck of the ten sails," and is a classic in marine annals. The lead ship of a group of ten merchantmen foundered on the shallow, windswept reefs off the eastern end of Grand Cayman. Despite signals warning off the other nine ships, all went aground. Heroic action by the Caymanians saved all of the crew and passengers of the ten ships, and not a single life was lost. A grateful English monarchy subsequently granted the Caymans' freedom from taxation in perpetuity.

This splendid absolution was the foundation of the Caymans' spectacular modern growth. With the development of modern bureaucratic nations tax burdens grew, and it was only a matter of time before the implications of "free of tax in perpetuity" attracted attention. The results were startling.

In recent years Grand Cayman has become a leading offshore financial center attracting an astonishing number of banks and corporations. Believe it or not, there were 16,712 corporations registered in the Caymans at the end of 1982. Considering that the 1982 population of the islands was 16,677, at that moment corporations outnumbered residents in the Caymans. In addition, 428 banks are licensed to conduct business in the Caymans; of these, 395 were licensed only to conduct business with nonresidents.

The flow of capital from this financial activity helped fund the development of tourist facilities in abundance. Hotels, condominiums, and diving services burgeoned. Frequent and reliable airline service completed the infrastructure, and the world's biggest diving boom ensued.

Throughout the recent dismantling of the British Empire the Cayman Islands, not surprisingly, have remained loyal subjects of the Crown. During the Falklands crisis, private funds appeals raised $500,000 to support England in its South Atlantic effort.

The Cayman Islands are located in the center of the Caribbean Sea along the edge of the 24,000-foot-deep (10 kilometers) Cayman Trench.

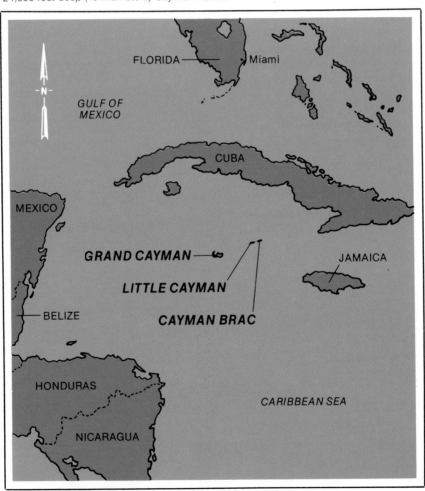

The Cayman Islands Today

The Caymans offer an abundant variety of tourist, diving, and financial services. 1982 saw over 5,000 airline flights into the islands, bringing nearly 120,000 passengers during the year. Most activity is centered in George Town, the capital, which dozes in the tropical sun at the southern end of Seven Mile Beach.

The beach itself, running northward from George Town to the rocky Ironshore, is an extravagant row of large and small hotels, condominiums, and restaurants. Other guest houses and hotels proliferate around the periphery of Grand Cayman, as well as on the smaller islands.

For divers, the result is the most complete array of choices in the world—snorkeling, beach diving, day boat diving, and live-aboard diving. Whatever you want, it's here. Shallow coral gardens, steep walls, wrecks—a diver's dream.

Weather. Weather is benign year-round. During the winter months some of the arctic waves die out this far south, and air temperatures can drop to the low 70s (about 20 degrees C). What is important for divers,

Snorkelers swimming in the shallow, clear waters off North Sound Reef.

however, is that even a cold wave has little or no effect on undersea conditions. There is always at least one area where divers are going at it, regardless of weather.

For most of the year, air temperatures are in the 80s (30–40 degrees C); water temperature was from 77 degrees F in winter to 82 degrees in summer (25–28 degrees C). Water visibility can fall to 60–80 feet (18–24 meters) laterally, but most of the year you can count on 100-foot (30-meter) visibility or better on the walls away from shore. Many do dive without any form of wetsuit; however, a lightweight one is recommended for comfort and protection against the corals. Also recommended are a pair of cotton gloves for protection against stinging hydroids.

When you visit be sure to watch your exposure to the sun. Many a vacation has been spoiled by early sunburn, and travelers are urged to exercise caution.

Hotels

Grand Cayman offers 15 different hotels and approximately 25 condominiums. The differences between the two are in the extent of services available. Grand Cayman hotels offer a variety of services to their guests, including meals, diving services, and watersports, to name a few. Condominiums are basically efficiency appartments with kitchen facilities. Luxury condos and others are offered on a daily or weekly basis and arrangements can be made with dive shop operators. Hotels charge for the rooms on a per guest basis, while the condominiums are rented by the apartment.

Ninety percent of Grand Cayman's hotels and condos are at the western end of the island near George Town, as are the restaurants. The western end of the island is also the location of Seven Mile Beach, which sports the largest concentration of hotel and resort facilities, such as sailing, snorkeling, scuba, and of course sunbathing and swimming.

Transportation

Hotel courtesy vans are not a common sight at the airport, as most of the hotels leave the driving to the taxi association. According to Cayman law, about the only vehicles allowed to make commercial passenger pickups at the airport are licensed taxis. The only exception to this rule is made for the hotels located at the opposite end of the island, at the East End or North Side. These resorts provide courtesy transportation in such cases because of the long distances.

Foreign Exchange, Dining, and Shopping

Currency. Two types of currency are in circulation on the Cayman Islands—U.S. dollars and the official Cayman Island currency, C.I. dollars. U.S. dollars are widely accepted, but if you pay for something in U.S. dollars you are likely to receive your change in C.I. dollars. Cayman paper notes are very colorful and are issued in denominations of $1(blue); $5(green); $10(red); $25(brown and orange); $40(purple); and $100(orange). Coins presently in circulation are in denominations of 1, 5, 10, and 25 cents. The C.I. dollar system is similar to that of the U.S.—one Cayman dollar is equal to 100 Cayman cents—but the two currencies are not equal in value. Although the exchange rate is fixed at C.I. $1.00 = U.S. $1.20, you may be able to exchange your dollars at a bank for a more favorable rate.

Shopping. The shopping opportunities on Grand Cayman are unique. There are items available here that are not found anywhere else in the Caribbean. Artisans of the island are experts in making black coral jewelry. There are at least a dozen stores offering a wide selection of hand crafted coral and other products. There are also elegant gifts from Europe; expensive perfumes, beautiful cut crystal, china, fine liquors, cameras, and electronic equipment are all offered at attractive prices.

Dining. Visitors to the Cayman Islands are not at a loss for dining experiences. Restaurants featuring everything from elegant gourmet dining to fast food can be found. The types of cuisine are as varied too; fresh seafood, lasagna, submarine sandwiches, even Chinese food are available. One word of caution, however: it would be wise to make reservations in advance, especially during the busy season, as many restaurants are booked solid.

Besides Diving. The Caymans are superb for diving, but when the diving appetite is sated, the islands are beautiful for walking, driving around to see the sights, or just lying about on the beach.

Customs and Immigrations

En route by plane to the Cayman Islands, you will be asked to fill out an immigration entry form. The law requires proof of citizenship, a completed entry form, and a return air ticket. A passport is not necessary, although it is the easiest form of identification. A birth certificate or voter registration card is also acceptable, but a driver's license is not. Customs inspectors are required to go through your suitcase, dive-gear bag, and whatever carry-on luggage you may have. Be forewarned that the Cayman Islands have the most severe anti-drug laws in the Caribbean. Violators are immediately arrested and taken to jail, and prosecuted violators can receive jail terms up to ten years and fines up to $20,000. Visiting tourists are not immune to the enforcement of these laws.

Except for some cliffs on Cayman Brac, the islands barely rise above sea level. George Town harbor is a busy cruise port, with large ships such as this one coming and going continuously.

2

Diving in Grand Cayman

The many popular reef sites of the Cayman Islands have common characteristics, so much so that certain types of dives are almost trademarks of
this island group. Because the dive sites are simply well-known points on
continuous reefs that surround the islands (for example, Trinity Caves, the
Tunnel, and Orange Canyon are merely different places to drop anchor
along a single quarter-mile segment of coral wall), you may easily swim
from one to another.

For this reason there is a certain artificiality in naming reef sites as if
they were separate and distinct. In truth, the Caymans have hundreds of
dive sites, because you can drop anchor anywhere along the miles of
dropoff and still find good diving.

Dive operators have become accustomed through the years to using
certain sites again and again. Thus, you can be aboard one boat anchored
off Seven Mile Beach and observe another boat anchored on the *Balboa*
wreck, another on the Aquarium, another on Peter's Reef, and so on. The
continuous nature of the reefs explains the feeling one has of a certain reef
profile being a "Cayman type of dive."

Indeed, whenever I think of Cayman my mind's eye soars vicariously
into this typical scene: We are anchored at the end of a long, gradual
bottom-slope from shore that extends anywhere from 300 to 900 feet, or
about 100 to 300 meters. As I look downward from the surface there is a
crowded bed of coral at a depth of 45–60 feet, 14–18 meters. Often this bed
of coral is crossed by channels running perpendicular to the shoreline.
Abruptly, this reef crest ends, and there is a sharp transition to deep water.
In some areas the reef crest is like the edge of a table, straight and sudden;
at other sites the edge of the coral crest is furrowed with deep channels or
gullies which extend downward to depths of 150 feet (45 meters) or more.

Because the reef crest is a distance from shore at most sites, the
Caymans have evolved as a boat-diving destination. With rare exceptions
the dive sites are simply too far from shore to swim out for diving.

*Eagle Ray Alley, a gash in the face of the North Wall almost 30 feet (10 meters) deep, was
named for the huge rays that are seen cruising the wall here.* ▶

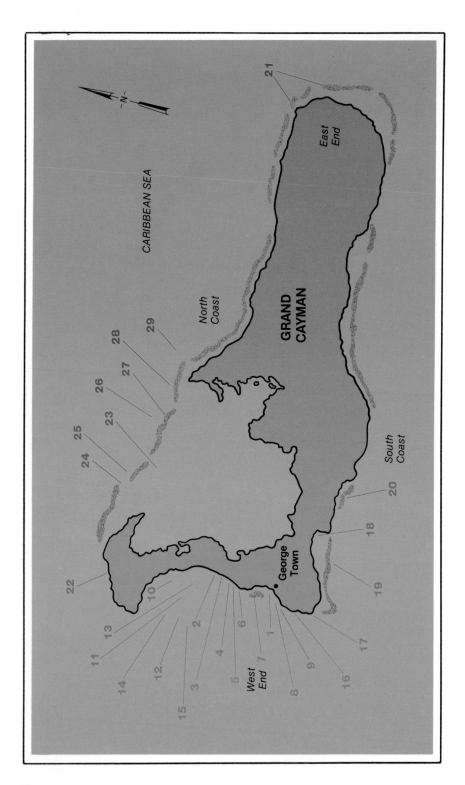

The three Cayman Islands—Grand Cayman, Little Cayman and Cayman Brac—are very similar in their topography both above and below water. All three offer grand scenery to the diver. Were each of these islands not part of this stunning chain, they would still be well worth visiting individually.

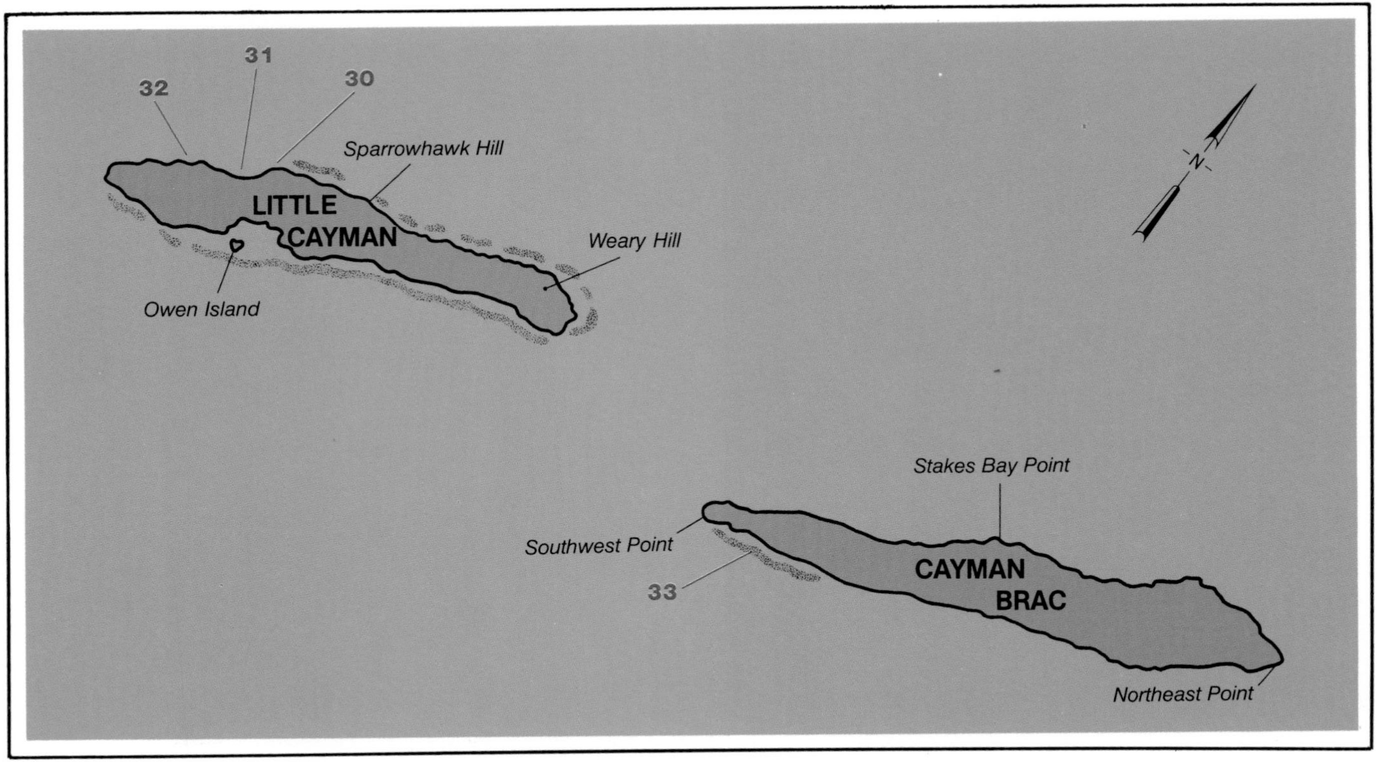

LITTLE CAYMAN

Weary Hill

Sparrowhawk Hill

Owen Island

30

31

32

CAYMAN BRAC

Stakes Bay Point

Northeast Point

Southwest Point

33

N

The deep precipices of Cayman's walls draw divers like a magnet. In these clear, calm waters, it's easy to forget how deep you really are. While diving Cayman, listen carefully to the information on each dive site given by the divemasters. Carefully monitor your depth gauge, watch, and pressure gauge during each dive.

When using the accompanying chart see the information on page 5 for an explanation of the diver rating system and site locations. ▶

Dive Site Ratings

	Novice Diver	Novice w/instructor or divemaster	Intermediate Diver	Intermediate w/instructor or divemaster	Advanced Diver	Advanced w/instructor or divemaster
West End						
1 *Balboa* Wreck*	×	×	×	×	×	×
2 Aquarium	×	×	×	×	×	×
3 Peter's Reef	×	×	×	×	×	×
4 Victoria House Reef	×	×	×	×	×	×
5 *Oro Verde* Wreck	×	×	×	×	×	×
6 Royal Palms Reef	×	×	×	×	×	×
7 Soto's Reef	×	×	×	×	×	×
8 Devil's Grotto*	×	×	×	×	×	×
9 Eden Rocks	×	×	×	×	×	×
10 Bonnie's Arch	×	×	×	×	×	×
11 Paul's Reef			×	×	×	×
Western Dropoff						
12 Trinity Caves			×	×	×	×
13 Orange Canyon			×	×	×	×
14 The Tunnel			×	×	×	×
15 Pinnacle Rock			×	×	×	×
16 Lambert's Cave			×	×	×	×
South Coast						
17 Original Tarpon Alley		×	×	×	×	×
18 South Sound Garden		×	×	×	×	×
19 South Sound Dropoff				×	×	×
20 Red Bay Gardens, Spot's Bay		×	×	×	×	×
East End						
21 East End Reefs		×	×	×	×	×
North Coast						
22 Hepp's Pipeline*	×	×	×	×	×	×
23 North Sound Anchorage*	×	×	×	×	×	×
24 New Tarpon Alley			×	×	×	×
25 Eagle Ray Alley			×	×	×	×
26 Pete's Ravine			×	×	×	×
27 No-Name Reef			×	×	×	×
28 Grand Canyon			×	×	×	×
29 Brinkley's Bay	×	×	×	×	×	×
Little Cayman/Cayman Brac						
30 Jackson Point	×	×	×	×	×	×
31 Bloody Bay Wall	×	×	×	×	×	×
32 Western Bloody Bay	×	×	×	×	×	×
33 Cayman Brac Reefs	×	×	×	×	×	×

* Indicates Good Snorkeling Spot.

West End

Soto's Reef 7

Typical depth range	:	5–35 feet (2–11 meters)
Typical current conditions	:	None
Expertise required	:	Novice or better
Access	:	Beach or boat

One of Grand Cayman's most popular reefs is Soto's Reef, lying just offshore from the Lobster Pot Restaurant. Here mound-like coral reefs rise from a white sand bottom at 35 feet (11 meters), with narrow corridors running between the coral mounds. Because it is so close to shore, this reef area has been dived heavily over the years, and it shows the impact of its popularity in wear and tear.

This reef runs from the Lobster Pot Restaurant northward to a point offshore from the old Pageant Beach Hotel. Naturally, there are several places to anchor , and some are even dubbed with separate names by enterprising boatmen. Fords Head, Ports of Call Reef, and so on are merely anchorages on different parts of the same coral complex. As you move toward the northern end of this reef, the large coral structures become more separated.

Serpulid worms, such as these attached to a star coral, and other small creatures make Soto's Reef an experience-building dive for novice macrophotographers.

Typical depth range	:	5–40 feet (2–13 meters)
Typical current conditions	:	None
Expertise required	:	Novice or better
Access	:	Beach or boat

Slightly to the southwest of Devil's Grotto and right along the same stretch of shore lies the coral mass known as Eden Rocks. The water is very shallow at the shoreline, and snorkelers and divers can enter from shore. It's a bit rough on the feet getting in, but once in the water the snorkel or dive is a breeze. The coral rises to within five feet (less than two meters) of the surface, so as at nearby Devil's Grotto snorkelers can see everything that's going on from the surface.

As a confirmed boat diver, I prefer to access this site from a boat approaching Eden Rocks from the seaward side.

Like Devil's Grotto, Eden Rocks shows the wear and tear of a lot of divers and is probably of interest principally to novices. Photographers, however, may find the schools of herring a rewarding subject for a warm-up dive before moving on to the deeper reefs.

A curious trumpetfish allows a diver to make a close inspection among mounds of staghorn coral near Eden Rocks.

Typical depth range	:	25–35 feet (8–11 meters)
Typical current conditions	:	None
Expertise required	:	Novice or better
Access	:	Boat

Almost directly out from the main pier of George Town lie the scattered remains of the freighter *Balboa*. Pieces of huge wreckage lie on white sand in an easy setting for divers. For this reason the *Balboa* is one of the most popular dive sites in the Caymans. It is also an excellent place for new divers, or divers coming back to the sea after a long absence. Finally, it is excellent for snorkelers, because all the elements of the wreck are visible from the surface.

Two high spots are not to be missed. The boiler is a huge cube with obvious boiler pipes filled with *Lima* file shells. A very large green moray has also been known to frequent the boiler room. The boiler has a large number of tame fish who feed on the invertebrate growth that coats its surface. On occasion, unusual creatures such as the orange frogfish *Antennarius* have been photographed here. The buoy which marks the location of the *Balboa* is tied to the boiler.

The second high spot is the propeller, found by swimming in a southeasterly direction from the boiler. The propeller and its mount rise from curving metal at a depth of 20 feet (7 meters). It is a graceful photo prop when you pose a diver with it.

With the exception of the boiler and the propeller, the rest of the *Balboa* is merely twisted wreckage. Since it lies in a shipping area, dynamite and other methods were reportedly used to lower its profile so boats could clear it.

Another treat during dives on the *Balboa* is a school of aggressive sergeant-majors. These small, swift fish have been fed many times, and at the merest hint of a plastic bag or plastic camera housing they move to the attack, swarming like a cloud of mosquitoes. Divers can literally disappear from view in this rolling mass of hungry fish.

Divers should explore far and wide on the *Balboa* wreck. There are flounder, eels, small groups of squirrelfish, angelfish, and other photo subjects everywhere. In some areas there are groups of sponge encrustation, clusters of serpulid and sabellid worms, brittle starfish, and other exciting small life.

The Balboa wreck in George Town Harbor has been almost completely flattened. Hordes of tame sergeant majors inhabit the wreckage, and large arrow crabs can be found inside the ship's boiler room. ▶

Night Diving on the Balboa

The *Balboa* site is the Caymans' most frequent night spot. Because it is an easy, shallow dive, a wide range of experience can be accommodated. The relative barrenness of the metal helps new divers spot octopus and other nocturnal creatures, and many a diver has had a first encounter with an octopus or moray eel here.

Typical depth range	:	30–40 feet (10–13 meters)
Typical current conditions	:	None
Expertise required	:	Novice or better
Access	:	Boat

Off Seven Mile Beach, due west from Harbor Heights condominium, lies a shallow reef known as the Aquarium. Although this reef is fairly close to shore rather than out along the main dropoff into deep water, it is far enough from shore to be a boat dive, inaccessible by swimming from the beach.

The Aquarium is not a separate reef area but a portion of a larger area that also includes Peter's Reef, which I'll describe later.

Marine Life. The Aquarium is named for its excellent fish population, in a serene reef setting 30 feet (10 meters) beneath the surface. The bottom is a series of shallow sandy gullies separated by modest ridges of coral. On the ridges you'll find clusters of antler coral, blue gorgonian fans,

Spotted trunkfish are extremely shy and hard to approach. One place to look for them is The Aquarium. Its shallow, inshore location makes it excellent for neophyte photographers or as a second dive after going deep on the outer wall.

Blue-striped grunts are common on all of the shallow reefs of Grand Cayman. Grunts are usually found swimming in groups. When disturbed, they will sometimes emit a distinctive, low-pitched grunt that can be clearly heard underwater.

and small sea whips. This invertebrate life provides minimal shelter for a number of goatfish, snappers, porgies, trunkfish, and brassy yellowtails. Wrasse and parrotfish abound.

This site is used by the boat captains in two circumstances: when they have a number of beginners aboard and want a shallow but entertaining reef, and after the divers have done a deep dive on the outer wall and the captain doesn't want them to accumulate any further decompression exposure. The Aquarium's location makes it a very sheltered, brilliantly lit underwater scene. While it is not as dramatic for underwater photography as the plunging outer walls or caverns, picture taking is very easy for new photographers because the scenery is well lit, the water is usually clear, and the fish are quite tame.

Peter's Reef 3

Typical depth range	:	30–40 feet (10–13 meters)
Typical current conditions	:	None
Expertise required	:	Novice or better
Access	:	Boat

Peter's Reef is located south of the Aquarium and the Harbor Heights condominium, about halfway between Harbor Heights and the Governor's home.

This is another site along the same reef complex that includes the Aquarium. However, it is distinguished by some extraordinary fish which have long been fed and entertained by Peter Milburn. Peter is one of the independent day boat owners and has worked this reef for years. He'll take along a mirror, lay it on the bottom, and a parrotfish will do pirouettes on its nose against the glass. Or perhaps he'll offer a fish the handle of his dive knife and the fish will mouth it like some playful puppy.

At Peter's Reef, a parrot fish finds a mirror, brought by the divers, a source of amusement. Damselfish are particularly intrigued by mirrors. These aggressive little tropicals think their reflection is another damselfish moving in on their territory, and will attack the reflection.

The beautiful black and yellow French angelfish is a mainstay of Cayman's Reefs. Found among the tangled barrows of staghorn coral atop the wall in West Bay, some grow to the size of a serving platter.

The terrain is composed of the same shallow sandy gullies and modest coral ridges that characterize the Aquarium. If you leave Peter's fish-feeding site and move toward deeper water, the reef thins out to open sand and scattered patch reefs at depths of 40 feet (13 meters). Far out across this open sand is the coral massif that heralds the main dropoff.

Like the Aquarium, this is an excellent beginner dive, or no-decompression spot. Photography is very easy; if you finally tire of the tame fish in the gullies there are two pairs of tame French angelfish out near the sand.

This is one of the rare shallow dive sites in the Caymans; it is therefore a good place for photographers to shoot those "reef scene with fish" shots in good ambient light. Out on the big dropoffs the reefs are deeper and the light levels are not as good.

Victoria House Reef 4

Typical depth range	:	30–45 feet (10–14 meters)
Typical current conditions	:	None
Expertise required	:	Novice or better
Access	:	Boat

This is the third in a series of shallow reefs which offer excellent second dives after a first dive on the deeper wall. This area is directly off the Silver Sands condominium, but as with the Aquarium and Peter's Reef, access is by boat and not by swimming from shore.

While the other reefs were practically level and punctuated by gullies, Victoria House Reef has more of a slope. Trailing off to 45 feet (14 meters), its corals thin out and end in open sand.

Still a shallow area, Victoria House Reef nonetheless supports larger examples of sponges and corals than the other in-shore reefs. Here, a brittle starfish hides inside a brilliant orange tube sponge.

The spotted moray eel is common in Cayman. Like other fish, eels on the more popular sites have become quite accustomed to divers, and may emerge from their hiding places in the reef in broad daylight.

Marine Life. The gully-and-ridge terrain is a bit more pronounced here, with the ridges somewhat wider and more massive than the modest ones of the sites mentioned previously. Corals, sponges, sea whips, and sea fans all are somewhat larger here. One would guess that the lazy, almost unnoticeable currents that move along the Caymans' coasts must eddy here and provide a richer plankton supply for the invertebrates.

All of these shallow sites share one characteristic—after thousands of divers, the fish are fearless. Porgies, parrotfish, and striped grunts hover before the camera, and even novice photographers can get good portraits here.

In your zeal for fish pictures, though, don't neglect the invertebrates. I've looked under the edges of barrel sponges or corals here and found octopus, lobster, spotted moray eels, arrow crabs, crinoids, and other superb subjects framed by the larger setting.

The *Oro Verde* Wreck   5

Typical depth range	:	25–50 feet (8–15 meters)
Typical current conditions	:	None
Expertise required	:	Novice or better
Access	:	Boat

This intact wreck of a small freighter was purposely sunk in May of 1980 to provide a new dive attraction. The first storm after the sinking promptly moved the wreck 300 feet (100 meters), leaving an impressive furrow in the white sand bottom on which it sits.

Located a quarter mile (about two fifths of a kilometer) off Seven Mile Beach in front of the Holiday Inn and Galleon Beach hotels, the *Oro Verde* will someday be a superb showpiece. In the course of time its currently

Once used by drug smugglers operating out of Miami, the Oro Verde *("green gold" in Spanish) now carries a cargo of marine life. After running aground in the Caymans, the freighter was bought by local divers and businesses. The vessel was cleaned and stripped of hatches and glass, and deliberately sunk in West Bay.*

drab metal should slowly encrust with corals and sponges and it will join such famous wrecks as the Truk Lagoon Fleet. The diving world's greatest wrecks need 20–40 years for marine life covering to really mature. There is one problem, of course—*Oro Verde* is dived so frequently that the corals and sponges may have difficulty surviving the normal wear and tear of a popular dive site.

Oro Verde is a fine novice dive, and a great opportunity for new photographers to get some good wreck shots. Unlike the *Balboa*, which was broken to wreckage because it was a navigational hazard, *Oro Verde* will remain intact until the sea itself consumes it.

A Treat for Divers

The *Oro Verde* ran aground after being stuck on a sandbar in 1976, and the 692 ton, 181 foot Panamanian cargo ship was then deliberately sunk 300 yards offshore for diving purposes. A group of Grand Cayman divers, with help from local hotels, tourist organizations, and airlines, purchased the salvage rights from the government of the Cayman Islands. The freighter was then stripped of its hatches, doors, glass, and other rough edges, making it as safe as possible for diving. The removal of its hatches and doors also creates a lovely sight as sunlight filters through the ship's interior.

A number of curious fish use the *Oro Verde* as their home, a phenomenon we find on wrecks around the world. Fish that know the labyrinth of a wreck are safer from predators than they would be on a reef because passing predators don't know the structure and find it strange, perhaps even threatening.

The *Oro Verde* lies tipped to port with its superstructure tilted away from shore. In the clear water, it provides a comfortable novice dive with easily followed landmarks.

Royal Palms Reef 6

Typical depth range	:	35–50 feet (11–15 meters)
Typical current conditions	:	None
Expertise required	:	Novice or better
Access	:	Boat

This relatively shallow reef lies offshore from the Royal Palms Hotel, southward from the *Oro Verde* wreck site.

This reef structure features a narrow cliff or crevice in the general shape of a crescent. The crevice has an extensive overhang, so that for part of its length it is almost covered like a tunnel.

Like most of the reefs in this series of shallow West End sites, it is a good place for novice divers. The terrain is a bit more developed than, say, the Aquarium or Peter's Reef, with more dramatic coral structures.

Royal Palms Reef, part of the same series of shallow reefs that includes the Aquarium, is home to a different group of marine creatures. The more developed coral structures here provide more crevices where small animals, such as this file shell, can hide from predators.

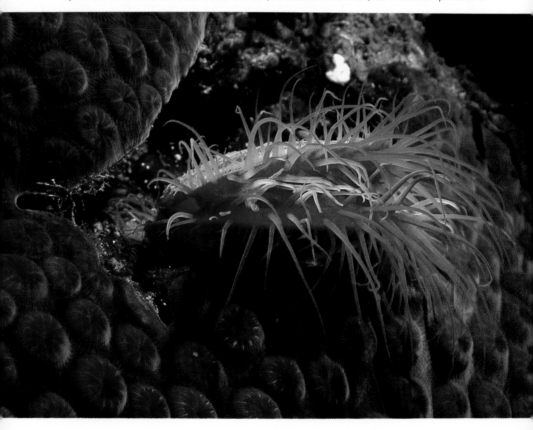

The coral overhang at Royal Palms Reef makes a perfect habitat for creatures that shun daylight. This bryozoan, a coral relative commonly called sea frost, can be found entwined among the branches of black coral growing in the shadows.

Another contrast with the shallow reefs is that an entirely different spectrum of creatures lives in the shadows of this type of reef structure. At a place like the Aquarium, for example, there is strong light penetration. You tend to find that nocturnal creatures try to avoid such areas, for they shun the light and prefer to spend their days secure in deeply shadowed crevices or under overhangs. But at the Royal Palms reef, anything from schools of dwarf herring to pink hydrocorals to big red crabs can be spotted in the shadows, and some black coral trees can be found thriving in the darkness.

At the bottom of the curving wall is a valley of white sand. Divers must be cautious not to kick up the sand with their fins, as the fine grains quickly fill the water with "snow" which can ruin your photographs.

It is the wall, the reef crest, and the life in the crevices that make this crescent a worthwhile dive.

Typical depth range	:	5–40 feet (2–13 meters)
Typical current conditions	:	None
Expertise required	:	Novice or better
Access	:	Beach or boat

This cluster of coral heads lies close to shore at the south end of George Town Harbor; looking directly out to sea you'd be looking over the site of the *Balboa* wreck.

Devil's Grotto is a smaller cousin to the well-known Eden Rocks, and is in fact directly offshore from the Eden Rocks Dive Shop.

This location is very convenient for snorkelers because it is close enough to shore to be an easy swim. Moreover, the corals come sufficiently near the surface that a snorkeler can see clearly the coral structure and any divers who happen to be working below.

For divers, this large coral complex is a honeycomb of crevices in which interesting invertebrates, small schools of dwarf herring, and other photo subjects are found.

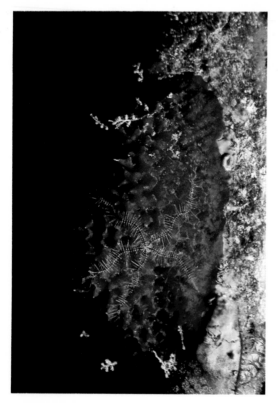

Devil's Grotto, easily reached from shore, is perfect for snorkelers or novice divers. Although worn down by the constant human traffic, it still holds a few surprises. Here, a brittle starfish moves across a blood-red sponge.

Blue tang, one of the more colorful of Cayman species, can be found grazing on corraline algae at many shallow dive sites.

Like any dive spot that is so easy to reach, Devil's Grotto tends to be a bit battle-scarred from a decade of heavy diving. For a novice diver, it combines easy access to shore with easy diving conditions, and is therefore an excellent place to start. For more experienced divers, however, I'd class it with Soto's Reef as a place to pass in favor of other spots further north or south.

Damage to the Reef. Those of us who have watched Grand Cayman develop into the world's most heavily dived island have mixed emotions. Even the most conservation-minded diver cannot help but damage corals with hands, knees, fins, or even kicked-up sand.

The damage done is cumulative and has begun to be most visible at the most accessible reefs. Another decade of this kind of wear and tear and sites such as Peter's Reef, Soto's Reef, and Devil's Grotto could be reduced to bare rock and sand. At the moment, however, they are the focal points of mass diving tourism, and vital to introducing new divers to the sea.

Typical depth range	:	35–60 feet (11–18 meters)
Typical current conditions	:	None to occasionally strong
Expertise required	:	Novice or better
Access	:	Boat

This unusual dive spot is in West Bay, at the northern end of the western coast. There are three separate attractions here. First are some rich coral gardens at 30 to 50 feet (10 to 15 meters) on rolling hills. These feature lots of well-developed corals, and as you putter about on the shallow corals you'll find trumpetfish, good-sized hogfish, and a marvelous array of tropical reef fish.

The second attraction is a small arch lined with corals and sponges. Divers may swim through the arch but should be careful not to exhale beneath it, as the rising bubbles not only dislodge debris but can damage the fragile corals. The arch is a beautiful spot for diver photos. In the small grotto beneath the arch, as well as on the small wall outside, there are some very attractive and photogenic sponges and gorgonians. One trick here is to avoid stirring up the bottom with your fins while you are shooting the arch; when you then drop to the bottom you don't want to encounter your very own sandstorm.

Bonnie's Arch, in West Bay, is an unusual formation. The fragile coral arch supports sponges, gorgonians and dozens of small fish species.

Near Bonnie's Arch, in the sand flats towards the wall, are a number of huge basket sponges. These sponges, some up to 8 feet (3 meters) in diameter, are one source of Cayman's deserved fame as a dive location.

The third major attraction of this site is the flat sandy bottom outside the arch. It has some good-sized basket sponges, wandering schools of blue tangs, French angelfish, and other fine photo subjects. There is one problem, however; after you've dived the arch, you'll have to watch your decompression exposure at this depth.

Cathedral Beauty

Bonnie's Arch is one of the finest dive spots on the entire West End. Though it lacks the mesmerizing impact of the deep dropoffs, its special charm is found both in the church-like grandeur of its arch formation and in the satisfying population of fish within 150 feet (30 meters) of the structure.

You may want multiple dives here, for there is a natural tendency to take most of your dives right at the arch, shooting wide-angle shots. You have to be sated with the arch before you can begin focusing on the smaller creatures all around you.

Typical depth range	:	50 feet (15 meters), 45-degree slope to unlimited dropoff
Typical current conditions	:	None
Expertise required	:	Intermediate or better
Access	:	Boat

Directly off West Bay point is a transition site between the West End Wall and the North End Wall. Paul's Reef is a long, 45-degree slope of coral which begins at depths of 60 feet (20 meters) and continues down to 200 feet (60 meters), at which point it drops vertically into deep water.

The boat usually anchors in on the sand, and the wind holds it out above the coral slope. The shallower portion of the reef is composed of coral similar to that of Trinity Caves and other areas a few hundred yards to the south.

As you proceed down the slope, you'll encounter some really huge barrel sponges at about the 100-foot (30 meter) level.

Strong Currents. This is one of very few dive spots on the main island which occasionally experience strong currents. Whichever boat you

Located between the sheer West End Wall and North End Wall, Paul's Reef is a long, deep slope. The coral gardens on the upper part of the slope harbor dozens of tropical species, such as this rock beauty, while the lower area is home to enormous barrel sponges.

*The currents that some-
times sweep Paul's Reef
can render the normally
crystalline Cayman waters
even more clear.
Deepwater gorgonians
and other filter feeders,
such as sponges, benefit
from the currents, growing
to exceptional size.*

use will certainly be alert to this potential for currents, and will either advise you before entry or even move to another site. Sometimes diving in current offers the finest in conditions for diving—the water can become crystal clear, and pelagic species such as schools of jacks may move in to feed on smaller plankton eaters.

Under good conditions this is a fine dive for photography, ranging from excellent macro and small-fish photography to the massive sponges and open-water creatures farther down the slope.

Decompression cautions. You will have to be cautious about your decompression time here. Long slopes in general mean that returning from the deeper part of the slope requires a long lateral swim.

For this reason, you'll want to leave the 100-foot (30 meter), level with plenty of air in your tank, so that you can follow the slope back to a point beneath the boat. You'll also want to begin your return well before the tables indicate you must; a sloping return is a slower return for decompression purposes.

After the sheer walls of the dropoff and the shallow gardens of the Aquarium and Peter's Reef, you'll find Paul's Reef an interesting hybrid and another facet of Cayman diving.

Typical depth range	:	45 feet (14 meters), to unlimited (wall)
Typical current conditions	:	None to moderate
Expertise required	:	Intermediate or better
Access	:	Boat

This is the first of a series of dive sites on the western dropoff moving southward from West Bay. Trinity Caves is directly off the point on the coast where the rocky Ironshore coast begins and the sandy beach ends.

Boats usually anchor at the edge of the sandy bottom that lies inside the reef mass. That avoids damaging the corals, as well as enabling the boats to retrieve their anchors easily.

At 45 feet (15 meters) you arrive atop the coral mass amid plentiful sea whips, antler coral, star coral, and sea fans, and even novices may enjoy themselves here.

Slicing through the coral mass here are a series of channels, through which one may swim with only occasional glimpses of the sky, for the upper edges of the channels are nearly grown together.

Following the channels, you are led directly out onto the West End's most spectacular wall. Just as you come to the wall there is an immense L-shaped canyon slashed through the solid coralline limestone. This is thrilling topography for divers making their first wall dives; the photographer can remain deep in this great slash through the coral and silhouette

At the end of the sand channels that lead to Trinity Caves is the West End Wall. Along its sheer face are deepwater gorgonians, sponges and black coral trees. Large open water species such as eagle rays and turtles cruise the cliff face, thrilling divers.

Like a Scottish road closely walled with piles of dirt and sod, coral mounds hem in the sand channels at Trinity Caves. Divers can play tag with French angels and other species which hover over the masses of staghorn coral atop the coral mounds.

a model against the electric blue of the open water. The crevice opening is so large it will dwarf the model for a stunning effect. The more advanced you become in your photography, the more you can use these complex settings for imaginative compositions of your own.

The Western Drop-off

There are lots of gorgonian fans, black corals, and sponges here for the photographers. For experienced divers there is the thrill of soaring out over the wall, watching the passing parade in the deep blue water. All along this wall you may encounter eagle rays, turtles, barracuda, and other larger animals. This is one of the West End's premier dive spots, and is always worth one more dive.

Some divers feel that diving various spots along a continuous wall must inevitably be boring. That's far from true. This wall, or the Great North Wall, can provide you literally hundreds of thrilling dives depending upon which animals or scenes strike you on any given day.

Typical depth range	:	45 feet (15 meters), to unlimited (wall)
Typical current conditions	:	None to moderate
Expertise required	:	Intermediate or better
Access	:	Boat

Orange Canyon is just south along the main western dropoff from Trinity Caves, offshore from the last buildings leading to West Bay Point. Its topography is similar to that of Trinity Caves, with a broad reeftop at a depth of 45 feet or 15 meters.

At the reef crest, furrows and ridges form colossal battlements that look out commandingly over a dropoff of several thousand feet. There are some big basket sponges here, some as shallow as 50 feet (15 meters) and others as deep as 150 feet (45 meters) and beyond.

The projecting ridges between the furrows are garlanded with characteristic dark gorgonian fans that can reach 4–6 feet (1–2 meters) in height. When there is a current running, these colonies will quiver in the clear water.

The "orange" of the site name is for several large orange sponges which nestle at 70 feet (21 meters) or so against one of the projecting ridges. Other sponges form fingers of orange, pink, purple, and red on the vertical face of the wall.

A scrawled filefish flees from an oncoming photographer along the wall at Orange Canyon.

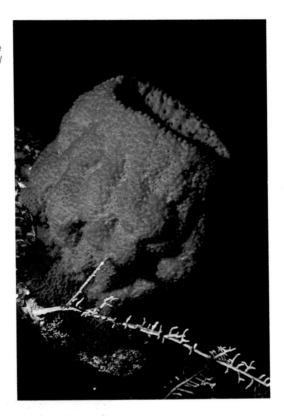

Several large orange sponges, such as this one, were the source of the name for Orange Canyon. The wall here is nearly vertical, and supports a variety of large sponges.

Triggerfish. On many dives here you may encounter schools of ocean triggerfish *(Canthidermis)*. These clumsy but charming fish are like large, awkward platters with fins on top and bottom. It is a strange design for a fish, peculiarly unsuited for either speed or grace in swimming. As you move down the wall these ungainly flappers accompany you at a respectful distance, soaring in great lopsided circles.

Other larger creatures may be spotted along the wall—barracuda, eagle rays, horse-eyed jacks, and turtles.

Other Marine Life. This western wall of Grand Cayman is one of the world's most impressive protected dropoffs, offering both easy access and dramatic topography. Although I am focusing on particular sites along the wall, you can find new treasures at any time merely by swimming along the reef crest looking for interesting things to explore. You'll find wandering angelfish, deep crevices slashing far back into the reef wall, huge sponges, outbursts of gorgonian fans, and a cornucopia of reef fish.

Seek and ye invariably find.

The Tunnel 14

Typical depth range	:	45 feet (15 meters), to unlimited
Typical current conditions	:	None to moderate
Expertise required	:	Intermediate or better
Access	:	Boat

The Tunnel is another site along the same western wall, southward from Orange Canyon. Once again the topography consists of a furrowed reeftop at 45 feet or 15 meters, which culminates in ridged battlements that drop off sharply into deep water.

The big attraction for experienced divers at this site is a pair of tunnels that slash downward through the reef mass to open on the wall itself.

The shallower of these tunnels is entered at 80 feet (25 meters) and if you follow it to its end you exit on the wall at 120 feet (37 meters). This tunnel is really large; three or four divers could actually swim abreast as they move through it. A second tunnel, which is specifically not recommended for sport divers, is entered at 160 feet (50 meters) and exits at 220 feet (67 meters).

The Tunnel is a cave which descends vertically through the lip of the West End Wall at 80 feet (28 meters) and exits on the face of the wall at 120 feet (40 meters).

Gray angelfish and other curious species around the Tunnel are quite used to people, and often follow divers. Because it's easy for divers to hide in the crevices just south of the Tunnel, this is also a good place to catch a glimpse of an eagle ray or other large pelagic creatures.

Marine Life. Like its neighboring reefs, the Tunnel reef boasts plentiful gorgonians, barrel sponges, and other invertebrate growth. There are also some quite tame French angelfish and gray angelfish that eye divers along the wall. We saw one gray angelfish circle the divers waiting for our exhalations, then rush in to bite the ascending bubbles.

Just to the south of the Tunnel, beyond an open valley on the wall, is a rather convoluted maze of twisting crevices leading upward toward the reeftop. There is a natural tendency to immerse yourself in among the creatures that grow in shadows of these crevices. While you are so occupied, an eagle ray or other bigger attraction could swim right by without your ever seeing it.

To avoid missing these when you are twisting through the tunnels, pause before exiting, take a breath, and look out to see what is passing by. This is often your best chance to photograph the larger animals precisely because your threatening bulk is hidden in shadow.

Typical depth range	:	Unlimited (wall)
Typical current conditions	:	None to moderate
Expertise required	:	Intermediate or better
Access	:	Boat

Pinnacle Rock is one of those sites on the western wall that has everything. The main formation is an immense ridge leading from the wall itself to a huge round rock. The rock stands out like a pinnacle from the surrounding coral mass, and is garlanded with photogenic sponges and gorgonians.

Along the southern side of the main ridge is a good-sized barrel sponge, which last time I looked sported an adult and juvenile of *Stenopus hispidus,* the barber-shop cleaner shrimp.

At the base of the great pinnacle are deep, shadowed crevices, and a small tunnel through the solid coral which is crowned by massive yellow-green tube sponges. This is a superb site for working with a model to shoot exciting diver shots.

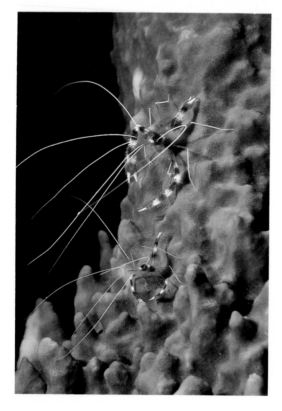

A pair of barbershop shrimp have set up a cleaning station inside this basket sponge near Pinnacle Rock. Remaining in one place, the shrimp attract larger fish, which hover motionless in the water while the shrimp crawl over them, removing small parasites.

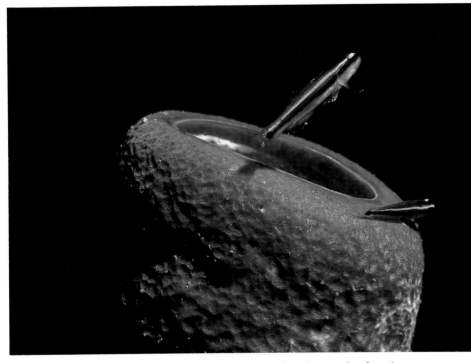

Gobies, such as these, are also cleaners—they remove and eat the parasites from the skins of larger fish. Gobies can often be found hovering near the end of a tube sponge, where they take shelter when threatened by a predator.

The vertical wall on three sides of the pinnacle is a thicket of gorgonians, sponges, black coral, clusters of bryozoans, and other rewarding photo subjects.

At each of these western wall sites (The Tunnel, Orange Canyon, Trinity Caves, and the Pinnacle Rock), a diver is on a compelling dropoff into deep water. Experienced divers, like experienced pilots, monitor depth, time, and remaining air supply constantly when working at depths approaching 100 feet (30 meters) or deeper.

Divers who are earlier in their careers can easily be lulled by clear water and the ease with which they can soar up and down these undersea mountainsides. Each diver must train himself or herself to monitor depth and decompression exposure habitually. You can be at 100-foot (30-meter) depth and be so comfortable that you fail to recognize crucial decompression limits.

For example, the dive on which I saw the two shrimp on the barrel sponge at 70 feet (21 meters) was an easy dive; I just sat motionless and photographed the tiny shrimp. Because this is such easy diving you might neglect the decompression tables—or concentrate so much on photography that you forget the passage of time. Be alert!

Typical depth range	:	50–125 feet (15–38 meters)
Typical current conditions	:	None
Expertise required	:	Intermediate or better
Access	:	Boat

South of George Town and generally off the Grand Old House is an interesting combination of coral and sand in a dramatic setting. Two massive coral complexes rise from a broad slope of white sand, with a chute or river of shadowed sandy bottom running downward between the heads.

The two environments offer totally contrasting types of diving. On the huge coral heads are gardens of coral gorgonian, sponge, and flashing reef fish. On one spectacular dive a huge black grouper challenged us right at the top of the southern coral mass. For five minutes my partner Jessica and I maneuvered with the big grouper and his escort of immature bar

This huge black grouper was found at Lambert's Cove. Although they ordinarily stay in deep water, the curious fish will sometimes ascend to within the depth range of sport divers. When annoyed, the groupers "boom" loudly, making a sound like a huge bass drum to frighten intruders.

Another resident of Lambert's Cove is the hogfish. Although they have been fished out in many shallow reef areas of the Caribbean, there are numerous varieties of hogfish around Cayman.

jacks. He watched us curiously, and at one point even settled down to eye Jessica face to face. It was a thrilling dive, and one we'll remember vividly every time we look at the pictures of this patriarch of the reef.

Groupers of that size often frequent deep water, staying at 200 feet (60 meters) or below and coming up into shallower reef areas only for food or out of curiosity.

Another denizen of Lambert's Cove is a huge hogfish or roosterfish. His preferred feeding ground is in the shadowed valley between the two coral masses. Here he buries his snout in the river of white sand and shell fragments, rooting for succulent tiny shellfish. Between snacks, he swims up to the diver as if to see whether he has any chance of a handout from the noisy intruder.

South Coast

Original Tarpon Alley   17

Typical depth range	:	20–45 feet (6–14 meters)
Typical current conditions	:	None to strong
Expertise required	:	Novice or better
Access	:	Boat

This dive area was first dived in the mid-1970s. I dived it from the original *Cayman Diver;* at that time a school of one hundred or more tarpon were almost always there, sweeping through series of shallow valleys and ledges. We would wander from valley to valley, never knowing when we would come face to face with these silvery, slant-jawed gamefish. When the tarpon came to us, we would swim with them over hill and dale. It was an incomparable experience. The tarpon were very patient, and would often swirl around us in a flashing-mirror display of what almost seemed affection, though we knew it was only tolerance.

Tarpon are swift predators and incomparable fighters when hooked by fishermen. Fortunately for the tarpon they are not good eating, so the

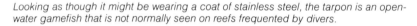

Looking as though it might be wearing a coat of stainless steel, the tarpon is an open-water gamefish that is not normally seen on reefs frequented by divers.

The rounded, broken blocks of coral on the floor at Tarpon Alley offer little protection from the strong currents that sometimes sweep the spot. Many of the tarpon who lived here in the past have moved to New Tarpon Alley, farther up the coast.

pressures on their population have been minor. Throughout the reef society the shadow of predation by man or other natural predators hangs over all but the bad-tasting. It's an argument for being born bony—or poisonous.

Another occasional visitor to these hills and valleys was a colossal sea bass named George, who was over six feet (2 meters) long and weighed several hundred pounds.

Alas, over the years an increasing traffic in divers gradually drove away most of the tarpon. When they are engaging in schooling or social behavior they are more tolerant of divers, but only to a point. That point passed, and while there are still some tarpon who visit these valleys, the principal population is now at New Tarpon Alley on the north coast.

Original Tarpon Alley is located some 300 feet (100 meters) off Southwest Cay, in one of the very few regions of occasional strong currents. Caution should be exercised. The bottom here is of rounded, broken blocks of coral, with small caves, natural arches, and winding canyons. When strong currents roar through it can be hard to find shelter; boatmen bringing you to this site will always check the status of the current for you.

Typical depth range	:	15–40 feet (5–13 meters)
Typical current conditions	:	None
Expertise required	:	Novice or better
Access	:	Boat

South Sound Gardens is an area of coral gardens some 300 feet (100 meters) from the wreck of the *Pallas*, off South Sound. It is near Pull and Be Damned Point. That catchy cognomen arose from local fishermen describing salvage attempts on the *Pallas*—they could pull, and pull, and be damned if they could get the ship off the reef. Today the bow of the ship juts above the surface, not far from some well-developed coral gardens.

There are massive elkhorn corals here atop a series of canyons, some of which are surmounted by natural bridges. The bases of the huge elkhorn colonies are at 20–25 feet (6–8 meters), while at the sand in the valleys you'll be diving at a mere 35–40 feet (11–13 meters).

There are also some wonderful shallow reef scenes on the coral mass; the elkhorn coral arms grope toward the surface, while schools of small reef fish swarm about them. I've come upon a dozen trunkfish socializing beneath a single huge elkhorn; at other times schools of snappers or Wrasse dance through the intricate garden-like scene.

The shallow waters of the Sound Sound offer some superb examples of elkhorn coral. When winds blow from the northeast, the south side of the island becomes a calm lee shore, offering exciting diving among massive elkhorn colonies.

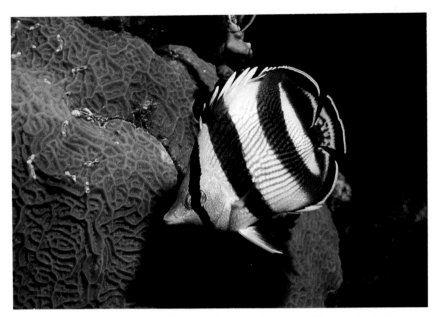

Banded butterflyfish can be found beneath the spreading arms of the stands of elkhorn coral in South Sound.

Winter winds are a plus.

Along this coast, weather can be a problem. Winds from the southeast or east can send rollers booming over these shallow reefs, making boat operations impossible. Even winds from the northeast may produce waves that wrap around the island and cause turbulence here. All that turbulence and shallow sunlit water, of course, have caused those massive Elkhorn colonies to grow so large.

Although diving on these south coast reefs is weather-dependent, when they are diveable they are truly spectacular. Some of the best diving here occurs when winter winds boom from the north; then this south coast becomes the lee of the island, and those few operators who come to the south coast can offer you superb diving on relatively untouched reefs.

Typical depth range : 70 feet (21 meters) to unlimited (wall)
Typical current conditions : None
Expertise required : Intermediate or better˙
Access : Boat

If you move seaward some 300 feet (100 meters) from the South Sound coral gardens, you reach the main dropoff which runs the length of the south coast. The profile of this dropoff is rather different from that of the north coast. Here, a broad expanse of white sand slopes from the shallow coral gardens to a series of formidable coral towers which rise 20–40 feet (6–13 meters) above the sand. When you soar down from your dive boat to the tops of these towers, you'll touch coral at a depth of 70 feet (21 meters). On the seaward side of the towers the reef drops abruptly for thousands of feet.

This precipice does not drop directly to the Cayman Trench (where depths reach 24,000 feet or almost 10 kilometers); the Trench proper is about 5 miles (8 kilometers) south of this spot. Still, from a diver's

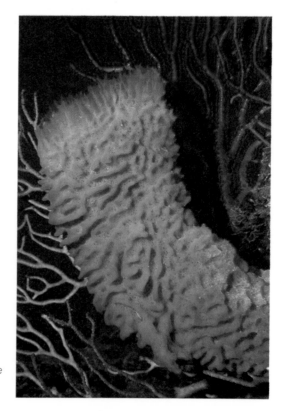

Another of Cayman's attractions is the azure vase sponge. Especially prolific here, the tips of the pink sponges appear to fluoresce with a pale, bluish-purple light.

The 24,000-foot-deep (7.3kilometers) Cayman Trench is just a short distance from the South Sound Dropoff. Open-water predators, such as the swift bar jack, cruise among man-sized tube sponges on this wall.

perspective it is a major wall. I've done dives in excess of 200 feet (61 meters) here, and it is straight down from there as far as the eye can see.

These massive walls are the real boundary from land to sea. Here the eagle rays, schools of jacks, turtles, and sharks live in a mostly empty inner space. Watching a pair of rays soar from darkness to darkness far down the sunless reef wall, we wonder what life would be like in that world.

Giant Sponges. Coral and sponge development along the precipice is extraordinary. There are man-sized tube sponges reaching out into the somber waters, and down the wall big basket sponges, gorgonians, wire corals, and black corals flourish. This is superb diving and well worth the wait for the weather to be just right. The *Cayman Diver II* and a few of the hotels come to this area any time they can. If you ask upon arrival, your dive guides usually can tell you if the weather is right.

The same cautions that apply to diving the North Wall apply here. Since the reef crest is deep, there is a tendency for divers to spend much of the dive at 70 feet (21 meters) or more. That requires close attention to decompression tables.

Typical depth range	:	15 feet (5 meters) to unlimited (wall)
Typical current conditions	:	None
Expertise required	:	Novice or better
Access	:	Boat

At least three more dive sites stand out along the south coast. They offer superb shallow gardens similar to those I described off South Sound. These are known as Red Bay Gardens, Spot's Bay, and an unnamed reef east of Spot's Bay off Charles Kirkconnell's home.

Red Bay Gardens is, not surprisingly, just offshore from Red Bay. It consists of sloping, shallow gardens ranging in depth from as shallow as 15 feet (5 meters) to as deep as 40 feet (13 meters). Like South Sound Gardens, these coral fields feature mature Elkhorn corals and a reef mass punctuated with twisting canyons and crevices.

South Bay Gardens, like other shallow areas along the south coast, features broad expanses of mature elkhorn coral. These gardens are rarely visited by divers, and are in much better condition than the shallow reefs of West Bay.

Red hinds are common in the crevices at the base of the large stands of elkhorn and staghorn coral along the South Coast. The very shallow depths, abundance of light and marine life make these areas a different kind of Caymanian treat for divers able to visit them.

Coral Fields. Spot's Bay offers some lush coral fields only 150 feet (50 meters) offshore from the cruise ship tourist dock at Spot's Bay. The reefs begin at Spot's and continue westward for some distance. Like Red Bay and South Sound Gardens, the Spot's Bay topography is shallowed, cleft with cracks and crevices and small caves. Elkhorn and Antler coral are common, and the reef fish population is plentiful and colorful.

The third of this group of dive areas is off Charles Kirkconnell's home. This is a really extensive field of intense coral growth, similar to the others mentioned here but substantially larger in area.

There is another, deeper site off the Lighthouse Club, halfway between Bodden Town and the East End. This is a reef which is both deeper (60–100 feet or 18–30 meters) and more involved, with larger caves, schools of sennets, and broad valleys where big sand tilefish build their intricately tunnelled homes.

Especially for beginners, these south coast reefs are really wonderful. Richer and more complex than the shallow reefs of the West End, they are protected by weather from the wear and tear of too many divers, too many hands and elbows and swimfins. For the few operations that offer them, they are a terrific bonus to the clients.

East End

East End Reefs 21

Typical depth range	:	30 feet (10 meters) to unlimited (wall)
Typical current conditions	:	None to strong
Expertise required	:	Novice or better
Access	:	Boat

The reefs of the East End are largely dominated by wind and wave, with weather playing a lead role in where and whether you dive.

There are a number of dive sites along this normally windward coast—High Rock Cave, Vertigo Wall, Grouper Grotto, and others. In many cases, the only way to dive these spots when wind prevails is by quick exits from small boats. The small boats have their divers suit up, then duck out through channels in the main reef and quickly get the divers over the side. This can be beyond the comfort level and competence of novices, and is best categorized as being for intermediate divers. Still, if the wind is completely down novices may enjoy these strong dive sites.

Although often subject to wind and weather, the East End dive areas offer some striking sights. Here, a tiny gobie is cleaning the parasites from the mouth of a tiger grouper.

The blood-red sponges of Cayman are justly famous. Found on the deeper reaches of walls on all sides of the island, they look dark purple, almost black, until a flashlight or photo strobe reveals their brilliant hue.

A Shark Watchers Paradise

One unusual attraction of these East End reefs is the presence of sharks. Throughout the world, larger animals such as big groupers, amberjack, and sharks are found in areas of rich feeding. These feeding sites are often promontories of rock and coral washed by heavy wave action and currents. Sites which display these conditions at least part of the time are excellent candidates for shark sightings.

The presence of sharks is *not* a reason for concern. The noisy thrashing and thunderous bubbles of a group of divers tend to send sharks packing. Indeed, it is hard to get a decent shark photo anywhere in the Caribbean unless you stage a feeding.

The East End reefs are dramatic in their topography, with chasms carved through the face of the wall, sizeable caves, overhangs, and slashing crevices. The excellent feeding from wave and currents that gathers larger swimming predators also means big sponges and rich coral development.

When the weather allows diving along this coast the diving is sensational.

North Coast

Before attempting to identify specifically the numerous dive locations on the long North Coast of Grand Cayman, some introductory remarks are necessary.

The North Coast extends for some 20 miles (about 32 kilometers), from the Turtle Farm in the west to Rogers Wreck Point at the eastern end. There are literally hundreds of dive sites along this coast, many more than could be accommodated in a guide of this size. Although several specific anchorages are used by dive boats whenever weather allows, the same captains often will simply drop anchor at a new spot in serene confidence that the diving is superb everywhere.

Another understandable difficulty is that captains jealously guard any good spot they find for their own clients. Thus, while areas such as Hepp's Pipeline, the Mini-Wall, North Sound Reef, the Grand Canyon, Eagle Ray Alley, and New Tarpon Alley are distinctive, the captain whose boat you use will have his own selection.

Two important factors will affect your North Coast diving. First is the weather; this coast is open to winter winds from the North and also to occasional brisk winds from the east and northeast. When these winds blow, boats at anchor above the North Wall experience quite a bit of rock and roll, making any diving operations difficult. In addition, visibility is quickly reduced when waves crashing on the reef stir sand and debris.

A second factor is that many hotels located along Seven Mile Beach on the West End find the North Coast too long a trip to service comfortably. Since the hotel boats return to their home bases for lunch, a commute of an hour or more to reach the North Wall can be too much to accommodate. This is especially true if the captain isn't sure the weather will be right when the boat arrives. Certain hotels, such as Spanish Cove and Cayman Kai, are located on the North Coast and naturally serve these reefs whenever possible. Also, the *Cayman Diver II* floating dive resort will take up residence on these reefs when the weather is right and stay in place for up to a week at a time.

The Cayman North Wall compares favorably with the world's greatest wall dives, and the fact that it is not dived as frequently as the crowded West End is one of its great strengths. The rays, turtles, and other pelagic animals have not been driven away by too frequent human incursions. Instead, because the weather makes these reefs unreachable for much of the year, these animals even seem curious about human visitors when we appear.

Typical depth range	:	25–60 feet (8–18 meters)
Typical current conditions	:	None
Expertise required	:	Novice
Access	:	Shore or boat

Hepp's Pipeline is one segment (the eastern end) of a quarter-mile (half-kilometer) long dive area offshore and slightly north of the Turtle Farm. In this area a sandy bottom with scattered corals leads outward from the shore to a depth of 25 feet (8 meters).

At this point there is a rim or reef crest of small coral heads with plentiful sea fans and small fish. Then the reef drops dramatically to a 60-foot (18 meter) sandy bottom. On this mini-wall, which is in some areas even undercut from the reef crest, divers find a multitude of photo subjects.

Along part of its length the face of the mini-wall snakes tortuously, presenting a rock profile that evokes the aerial view of a winding river. The stony face plunges and warps to its white sand bottom at 60–70 feet (18–21 meters), where blue tang, snappers, Nassau groupers, and even black groupers dart in and out of coral crevices just off the sand.

Here at the base of the mini-wall diver/photographers must exercise caution, as a stray swimfin can provoke a sandstorm.

The mini-wall faces north, so it is perpetually in shadow, and plentiful antipatharians (black corals, wire corals) grow there. There are also some substantial basket sponges, including one that is large enough to accommodate two divers for picture-taking.

This wall is both easy to dive and an extremely rich site. For this reason it is well suited to either macro or small-life photography. For fish photographers it is particularly rewarding, with an extraordinary population of colorful reef fish of all types.

Because of its quarter-mile length, this mini-wall offers several different dives all by itself. Dive boats often use it as a shallow dive after a dive on the major wall. Since the reeftop is so shallow, divers who have used up their no-decompression limits can still photograph fish and corals in depths of 15–25 feet (5–8 meters).

Some sections of the mini-wall offer surprisingly good diving, even for those fierce partisans of the North Wall. This shallow dive site has probably saved many divers from the siren call of the deeper diving along the north coast.

Snorkelers find easy access here too, though on occasion the water is cloudy and visibility from the surface may be reduced. It is one of the few places on the north coast where snorkelers can easily see scuba divers throughout their dives.

Typical depth range   :   Surface to 15 feet (5 meters)
Typical current conditions :   None
Expertise required   :   Novice or better
Access   :   Boat

Elkhorn Coral. This is a dive area that is full of charm, and is a wonderful spot for both snorkeling and night diving. The reef is composed of fast-growing elkhorn corals *(Acropora palmata)* which nearly reach the surface. The reef-line runs directly across the broad center of North Sound's opening to the sea. In aerial photographs it stands out as a solid barrier of coral across the central opening of the North Sound into the open sea. In some places along this wall, the elkhorn corals reach so near the surface a diver couldn't even swim over them to reach the other side. At either end of the coral barrier is a pass into the North Sound; a variety of fishing and pleasure boats leave and enter the North Sound through these channels. They are also the main entry and exit for rays entering or leaving the rich feeding on the shallow sand flats in the Sound.

I first experienced this reef when the *Cayman Diver* would anchor behind it for the night after a day diving the North Wall. Not only did it offer a wonderful night's sleep on flat calm water; night divers would return with tales of rays, turtles, crabs and unusual creatures of the sand.

As on the south coast, many shallow areas of the North Sound are crowded with thickets of elkhorn coral. The dense forests of coral, which almost reach the surface in some places, are separated by broad sand flats leading to open water.

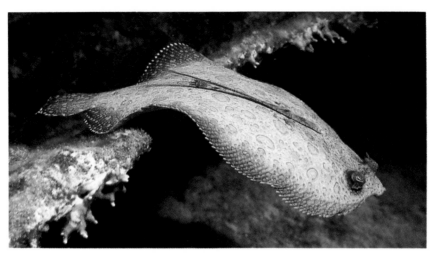

To confound predators, this peacock flounder can change colors at will, blending with the pebbly sand bottom.

The great thicket of corals that form the reef line are like an over-crowded gallery of sculpture, one great stone-armed creature crowding against the next. The massive thicket is a shelter to a myriad of animals, from schools of snapper to trunkfish to all types of invertebrates. Indeed, the broad sand flats around it offer no shelter at all, so the life concentrates along the reef line.

Shallow Water Hazards

Since the water is so shallow, divers should be reminded to use enough weights to stay down. There is a tendency to use the same weights as in diving the deeper wall, and you could find yourself too buoyant in this shallow water.

Another caution I would offer is to stay behind the coral line and not to wander out into the channels at either end. It can be very disconcerting to have a fishing boat suddenly churn overhead, its propellor slashing the water to foam close to your head. The channels aren't deep, perhaps only 15–20 feet (5–7 meters); they offer very little margin for safety.

Since the marine life concentrates in the coral, and the boats avoid it, divers are best served by working right along the coral line.

Typical depth range	:	55 feet (17 meters) to unlimited (wall)
Typical current conditions	:	None
Expertise required	:	Intermediate or better
Access	:	Boat

The Great North Wall, off the North Sound reef line, offers some of the Caribbean's most consistently exciting diving. The bottom slopes outward gradually from the North Sound reef line to depths of 50 to 70 feet (16–21 meters). There are some spur-and-groove formations, and some patch reef bottom. In some areas, such as Tarpon Alley, there is gleaming white sand bottom leading to a well-developed coral reef bastion, then a sudden

Although the large schools of tarpon have left Tarpon Alley on the southwest coast, similar schools can now be found at New Tarpon Alley on the north coast.

Rays are a Regular Sight.

Eagle rays are seen here regularly, and a six-foot (2-meter) long solitary hammerhead shark has also visited on numerous occasions. Scattered on the white sand behind the reef barrier are small patch reefs. You'll find southern sting rays, sometimes big ones, perched on the coral heads or browsing for crustaceans in the white sand. Sometimes visibility can be reduced due to tidal flows out of North Sound. While this is difficult for photographers, the animal life revels in the rich outpouring of planktonic and larval food.

dropoff of major proportions. Deep canyons slice completely through the coral rampart perpendicular to the reef face; this means that good-sized fish can simply disappear into these crevices and not be seen by predators. By the time they emerge at the other end of the crevice, any predator has completely lost track of them. These particular crevice formations dominate the reef crest here at New Tarpon Alley, Pete's Ravine, Eagle Ray Alley, No-Name Reef, and a hundred anchorages between them.

The outer face of the wall is gullied, so you dive a series of tall, jutting promontories separated by deep gashes into the reef face. Because the corals need access to passing plankton for sustenance, the best corals and sponges tend to be on the promontories. The valleys between are often empty sand or coral skeleton.

If you follow the promontories downward, the valleys even out at depths of 150 to 175 feet (45–53 meters), below which there is a shelf. Below the shelf the coral wall drops stunningly to depths of several thousand feet.

New Tarpon Alley got its name, naturally, from a school of tarpon which flock among the coral heads at the crest of the reef. The tarpon sweep in from the white sand behind the reef and flow through 20-foot (6 meter) crevices in the coral in a silvery, fluid mass. They are not agitated, and almost remind me of a school of dwarf herring flowing coherently about us, without isolating themselves in fear of being attacked.

The "New" part of the name (as you've already guessed) signifies the existence of an earlier Tarpon gathering place off the Southwest Point of the island.

Eagle Ray Alley

Typical depth range	:	70 feet (21 meters) to unlimited (wall)
Typical current conditions	:	None
Expertise required	:	Intermediate or better
Access	:	Boat

Eastward from Tarpon Alley, along the same reef crest and directly off the North Sound Reef, lies Eagle Ray Alley. As mentioned earlier, this wall offers a dramatic plunge into deep water all along this section of the coast. For most of its length it offers great diving, but certain sites have been singled out by boat captains to be visited again and again.

Eagle rays, so called because of their size and the majestic sweep of their wings, are commonly seen as they move in to feed on crustaceans on the sandy bottom of North Sound.

Eagle Rays. Dive sites usually draw their names from physical characteristics ("Grand Canyon," for example) or from exciting sightings of animals (Tarpon Alley). This spot regularly offers sightings of eagle rays cruising off the reef crest. I've seen as many as four together gliding silently from darkness to darkness, and there are a few creatures on earth as majestic and serene. The "Alley" portion of the name denotes a spectacular slash through the reef mass. This particular canyon, from open sea to inner sand flats, is 30 feet (10 meters) deep through solid coral and wide enough for several divers to swim abreast. While I haven't seen rays actually pass through the alley, I have no doubt that they do it regularly.

The reasons for the rays' presence are really quite mundane: they simply shuttle back and forth between the two sandy areas outside the channels into North Sound. Almost certainly they enter the broad, sand-bottomed sound itself to feed on crustaceans. Whatever the reason for their presence, they are a thrilling sight for any diver who encounters them in their peregrinations.

The promontory-and-valley pattern of reef crest dominates here; divers can photograph reef fish, corals, sponges comfortably on the rounded tops of the promontories without ever dropping over the edge.

Photographic Delights

If you do soar out into inner space, be on the lookout for some of Cayman's most spectacular photo subjects. At depths of 125 feet (38 meters) or more there are exquisite tube sponges in blood-red or deep orange. About the size of a silver chalice, these sponges assault your senses when you shine a hand-light on them. In that world of monochrome twilight, their presence is tremendously exciting, and some emotional photographers have been known to shoot an entire roll of film only on them.

Some of the sponges are decorated with dainty brittle starfish for an even more dramatic picture. Before you get carried away, though, watch for some spectacular gorgonians you can silhouette against the sun. There are also some immense tube sponges in purple and tan which are also very tempting. Cayman's reputation as the sponge capital of the world is secure.

Typical depth range	:	70 feet (21 meters) to unlimited (wall)
Typical current conditions	:	None
Expertise required	:	Intermediate or better
Access	:	Boat

Pete's Ravine is another dive site marked by a single undersea formation along the wall. In this case the ravine is precisely that, a ravine between coral promontories that is larger and deeper than others along the reef crest.

The promontories themselves are pronounced here. For a convenient visualization of the reef crest, hold your hand out before you, palm downward, and make a fist. Your protruding knuckles and fingers resemble the promontories and canyons along the North Wall. There are some deeply cut, shadowed canyons between the "knuckles" at Pete's Ravine, and long slender fingers of sponge hang down from rocky ledges with dazzling blue open water between them.

Another deep cut in the face of the North Wall is Pete's Ravine. Dangling rope sponges frame a diver hanging over the blue-black abyss beyond the wall.

The North Wall hosts many of the island's gaudiest sponges. This Chinese red cluster is nearly three feet (one meter) tall.

For photographers, these are some unusual effect shots to be essayed here. With the late morning sun rising above the promontories, you can look upward and see sun rays spraying out around the rounded pinnacle shape. Simply capture one or more tiny divers dwarfed by this incredible effect and you have an unforgettable memento of this area.

Pete's Ravine, like Eagle Ray Alley, New Tarpon Alley, and other spots, is regularly visited by eagle rays. Don't get too involved in photographing a sponge up close, for sure enough that's the moment a ray will soar past to look you over. When you're diving these areas, be ever ready with your wide-angle lens. Sometimes you have only a second or two to set your f-stop, be sure your strobe light is on, aim, and fire. Life's darkest moment can be the receding spotted back of a graceful ray you never saw coming.

Good Macro Photos. To make life more complicated, there are some stunning close-up subjects along the North Wall. In particular, indigo hamlets, diamond blennies in their host anemones, and tube sponges in blood-red, orange, and iridescent blue are common. Only for the sponges must you descend to 100 feet (30 meters) or more. The indigo hamlets and anemones are often right at the end of the promontories in plain view.

69

Typical depth range	:	50 feet (16 meters) to unlimited (wall)
Typical current conditions	:	None
Expertise required	:	Intermediate or better
Access	:	Boat

No-Name Reef is found along the reef crest off Rum Point channel, a half-mile (or about a kilometer) west of Rum Point. No-Name is distinguished by rather shallower promontory tops than those of its cousins further west. Where New Tarpon Alley, Pete's Ravine, and Eagle Ray Alley have you at 60-foot (18 meter) depths or more throughout your dives, here you can thread your way amid the promontories at 50–55 feet (16–17 meters) and enjoy a dive of longer duration.

At the aforementioned reefs, the reef barrier rises strongly from areas of white sand; at No-Name, the sand is merely in long gullies leading shoreward between the promontories. I can remember hanging off the anchor rope, thinking how perfect an example of spur and groove reefs this was, with its long slender white fingers of sand stretching away as far as the eye can see. As you near the reef crest, however, the sandy valleys give way to deep, slender crevices.

The reef top is a modest garden along the North Coast dropoff. Since the reef crest is 50 feet (16 meters) deep or more, the corals which dominate are tangles of antler coral, blue gorgonian fans, and some clus-

Azure vase sponges, glowing faintly blue at their tips, are among the attractions at No Name Reef on the North Wall. Similar to the other North Wall sites, No Name is a little shallower, allowing divers more time to explore the crevices and promontories.

Though the grouper which once infested this area have been fished out, plenty of tropical species remain. The edges of coral rock are frequently marked by the gouges of parrotfish. Using their tough beaks, these colorful reef inhabitants scrape algae off of the coral, allowing the coral to grow unmolested.

ters of iridescent blue vase sponges. Here and there along a promontory will arise an unexpected dome coral of several hundred pounds (100 kilograms or more), often decorated with blue gorgonian fans or small sea whips.

Tropical Fish. Unfortunately, generations of line fishermen have hooked many of the groupers along this wall, which must once have been grouper heaven. Today the fish that are left are those which won't rise to bait—squirrelfish, brightly-colored parrotfish, tangs, trumpetfish, all the journeymen of tropical reef species.

There is a great temptation to soar into the darkness far down the reef wall here. It's easy when the sun is high and the water clear. I've had small groups of horse-eye jacks circle me at depths of 100 feet (30 meters) or more, sweeping in from the distance in an unhurried manner, patiently watching me, confident in their sudden speed and grace.

It is always with great reluctance that I rise along the anchor rope after a dive along this wall. I always half-expect to see a manta or turtle or other spectacle just when it's too late—when the cameras are empty or the air is gone or the no-decompression time is used up.

No matter how many years I dive I'll never tire of roaming these great battlements; at their best they rival any dive anywhere in the undersea world.

Typical depth range	:	70 feet (21 meters) to unlimited (wall)
Typical current conditions	:	None
Expertise required	:	Intermediate or better
Access	:	Boat

Grand Canyon is a single formation, an immense cave-in of reef which has left a canyon 150 feet (45 meters) wide. On either side of this great canyon of rubble are immense coral buttresses jutting outward into crystal water.

This canyon is but one of several formations and dive sites near the Rum Point channel into the North Sound. While it is a spectacular dive spot, the other formations offer a variety of coral pinnacles, tunnels, and caves; this general area really offers a dozen superb dives. What all of them share, despite reef-crest differences, is a plunging, thrilling, mountainous wall. The gigantic coral buttresses normally mark the best diving; that is where the corals and sponges achieve their richest feeding. The buttresses are also where the angle of the dropoff from the reef crest is often perpendicular. Using your buoyancy compensator as your personal eleva-

With the sun directly overhead, photographers can silhouette divers perfectly against the massive coral escarpments at Grand Canyon. Of all the North Coast areas, this one offers the largest and most dramatic ravine scenery.

The bright blue indigo hamlet is a common inhabitant of Cayman reefs. Although the temptation on the North Wall is to go deeper and deeper in the still, clear water, some of the most beautiful species and lush growth can be found along the top of the wall.

tor, you can rise and fall with the coral in effortless diving. You can plunge 100 feet (30 meters) straight down the wall before it flares out to a second crest below 200 feet (60 meters). Then at a depth of 225–270 feet (70–83 meters) is a second dropoff that is sheer—and seems bottomless.

Like similar areas to the west near the North Sound Channel, the great cautions for divers concern depth. The immense wall beckons you deeper with a siren song. I've been with divers who, when they reached 275 feet (84 meters), thought they were "only" at 200 feet (60 meters) because of the effortlessness of diving these vertical precipices.

Even worse, the fact that the reef crest is at 70 feet (21 meters) means that any decompression must be done while hanging off an anchor rope. All of this can be avoided by well-planned, carefully executed diving. Please don't be lulled by the ease of diving deep on Cayman's North Wall. Sample the excitement, but exercise extreme caution. The few hotels and the floating dive resort which serve this area will reiterate these warnings strongly.

Remember, these are dives you will want to make again and again, so dive them safely!

Typical depth range	:	35 feet (11 meters) to unlimited (wall)
Typical current conditions	:	None
Expertise required	:	Novice or better
Access	:	Boat

At the eastern end of Brinkley's Bay lies a dive spot with the shallowest crest to be found along the entire north coast. This reef crest is a rather narrow edifice, with craggy battlements brooding above the deep-water dropoff.

Interestingly, the deep diving is not the only attraction here. Behind the ramparts of the reef crest lies a series of shallow sandy "streaks"— hardly wide or deep enough to be called valleys. A mere 60 feet (18 meters) across this streaked plain you'll find a raised garden of shallow corals— scattered waxy plates of fire coral, sea fans, sea whips, and a treasure trove of shallow water invertebrates.

This seldom dived area is a long way from the hotels of the West End, and so has been protected from the major traffic of divers on the island.

Mustard yellow or brown, fire coral isn't coral at all, but a relative of the jellyfish. The fine filaments extended from this fire coral are its stinging cells, which can cause a burning sensation if brushed with bare skin.

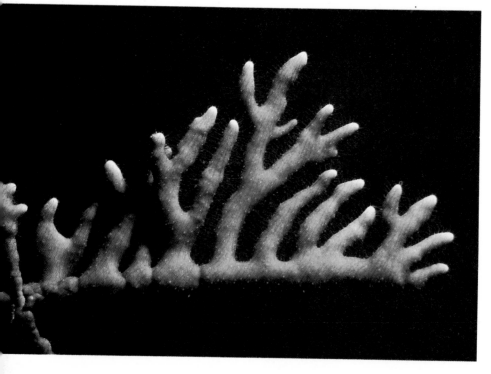

French angelfish are among the dozens of species that inhabit the Caribbean.

Little Cayman/Cayman Brac

Jackson Point 30

Typical depth range	:	30 feet (10 meters) to unlimited (wall)
Typical current conditions	:	None to moderate
Expertise required	:	Novice or better
Access	:	Boat

The reef off Jackson's Point is an interesting mixture of dive milieus. Inside the reef barrier is a broad, sandy plain upon which you will often find feeding or resting sting rays. These placid creatures usually become alarmed when approached too closely, but on occasion you'll discover one so busy rooting tasty crustaceans out of the sand that it will pose for a portrait.

As you move from the sand to the reef wall, there is an abrupt transition. The corals grow in a solid mass 60–90 feet (18–27 meters) thick, and the crest rises some 15 feet (5 meters) above the sand. Here and there are crevices through which you may enter the reef mass and swim in passages that penetrate through to the main dropoff. Along the inner rim of this coral barrier is propped an immense anchor, perhaps 10 feet (3 meters) tall, with a huge ring at the end of the shank. The anchor has been embraced by the coral, which has grown about it over the years. Much of its length is visible, however. Two other anchors are embedded in the reef and smothered in coral as well.

Wall Diving. Like the dropoffs of the North Wall of Grand Cayman, the North Wall of Little Cayman is a sheer extravaganza. From a crest at 40 feet (13 meters) below the surface, the wall plunges to beyond 180 feet (55 meters). A shallow ledge at that depth supports some hefty sponges, long strands of wire coral, and gorgonians. The wall then plunges to the abyss. Indeed, if you have dived the North Wall of Grand Cayman, just imagine the same awesome undersea topography but with a much shallower reef crest. As you can imagine, it is a stunning experience. The Little Cayman coast is weathered out even more frequently than Grand Cayman's North Wall, and the rays, turtles, and other dramatic life are even more plentiful here.

Like many Little Cayman sites, the vertical wall at Jackson Point is less than 30 feet (10 meters) from the surface. Azure vase sponges, tube sponges, enormous basket sponges and black coral can be viewed at depths of less than 60 feet. ▶

On many occasions, huge groupers, rays, turtles, and other pelagic visitors have overjoyed divers along the Little Cayman Wall. On some dives we've enjoyed a half-dozen sting and eagle rays in view all at one time. I've seen divers so excited by the diving here that they have made eight tank dives in a single day.

Without question, Little Cayman offers some of the finest wall diving in the Caribbean. It is extremely weather dependent, however, and frequently cannot be dived because of wind and wave action. When the weather is right, though, it's hard to find more spectacular diving in the entire Caribbean.

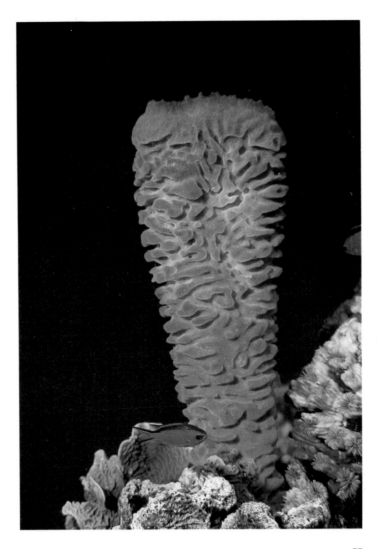

Typical depth range	:	20 feet (6 meters) to unlimited (wall)
Typical current conditions	:	None
Expertise required	:	Novice or better
Access	:	Boat

The Bloody Bay Wall curves westward from Jackson Point in a long arc. The reef crest along this great coral bastion is craggy and uneven, with numerous sand channels, caves, crevices, and tunnels; it offers some of the most interesting undersea topography in all the Caymans and is endlessly fascinating.

Scattered across the gleaming white sand plain behind the massive reef crest are smaller coral cities which rise to within 20 or even 15 feet (5 or 6 meters) of the surface.

The main dropoff is furrowed and sheer, making Bloody Bay another site where deeper diving is done merely by inflating and deflating one's buoyancy compensator. In preparing this book, I reminisced with Paul

As at Jackson Point, Bloody Bay is a vertical escarpment within hailing distance of the island's shore. Bright sponges and deepwater gorgonians can be found within easy sport diving depths, affording photographers enough time to shoot two or even three rolls of film before running out of no-decompression bottom time.

Bloody Bay is also a great spot for snorkeling between dives. Shoreward of the wall is a maze of coral canyons in just 20 feet (6 meters) of water. Filled with parrot fish, such as this colorful specimen, bar jacks, squirrel fish and schoolmasters, the area is a delight with or without a tank.

Humann, who was captain of the *Cayman Diver* for eight years. He reminded me of a dive four of us did to 275 feet (84 meters) on a very calm day; Paul reported being able to see us throughout the entire dive! Now *that* is visibility.

The sheer wall is slashed with canyons, some quite deeply etched into the precipice. Tangles of gorgonians, clusters of huge tube sponges, and black coral trees are found as you move deeper down the face. At 240 feet (24 meters) the wall slopes out to a ledge, then undercuts sharply and plunges out of sight in mind-bending darkness. At these depths one can be seized by sudden impulses—I can remember thinking of hanging on to the ledge to keep from falling into that endless night below.

Decompression Time. Another wonderful feature of diving this long table-edge dropoff is that you can dive down to 100-foot (30-meter) depths or beyond, then return to placid, gentle reefs at 15–20 feet (5–6 meters) for a bit of decompression time. These shallow reefs are quite rich at the reef crest, although they thin out to scattered small coral heads as you move 100 feet (30 meters) or so back from the crest. There are many Nassau groupers, French and gray angelfish, colorful parrotfish, and other tropicals in abundance, so that you can do extensive close-up photography while you do a bit of decompression time after diving the wall.

Shallow or deep, this is sensational diving; for any experience level, for any photographic interest, this ranks with the greatest.

Typical depth range	:	20 feet (6 meters) to unlimited (wall)
Typical current conditions	:	None
Expertise required	:	Novice or better
Access	:	Boat

The classic of all Little Cayman dives is to be found at the western end of Bloody Bay. Instead of chunky pinnacles at the edge of the dropoff, here we find a shallow table-land of corals—there are man-high gorgonians, small clustered dome corals where fish seek shelter, and waist-high sea whips everywhere. Nassau groupers abound here, ghosting from shelter to shelter like Indians moving through the forest. Their curiosity always betrays them, however, for they invariably come too close and you can't help seeing them lying amid the sea whips.

This long, flat expanse is 20–25 feet (6–8 meters) deep, with some coral heads reaching within 10–15 feet (3–5 meters) of the surface. This shallow reef top is invaluable after a dive down the nearby deep wall. Rather than hang on a rope, divers taking precautionary decompression may continue taking photographs amid these rich shallows.

The western end of Bloody Bay is, if possible, even more spectacular than the eastern end. Here, patch reefs as shallow as ten feet (3 meters) shelter Nassau groupers and other coral garden species. Mere yards away, the wall drops straight away, hundreds of feet (more than 30 meters), into blackness.

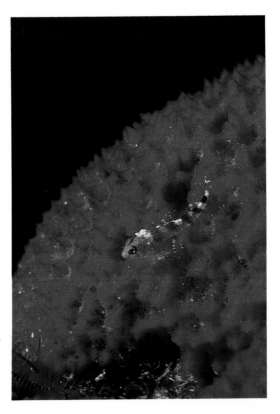

Blennies can often be found resting on the surface of the bright blood-red sponges along the Bloody Bay wall. A bit deeper, along a solitary ledge, is a grouping of brilliant orange elephant ear sponges, some six feet (two meters) across.

And what of that wall? Well, it's breathtaking. Like the edge of a table it shears off suddenly, masked only by some thin tendrils of gorgonian and stubby corals. Then you are suddenly over the edge, looking straight down.

The sheer wall bows slightly outward like a massive pot belly, and as your bubbles rise from 100 feet (30 meters) they will scour the bellying wall and dislodge debris. This is something for photographers to be wary of.

Sponges. At 180 feet (55 meters) there are some six-foot (two-meter) elephant-ear sponges in bright orange, just above the only ledge on the entire wall. The ledge extends outward at a 60 degree slope for only 20 feet (6 meters) or so, then plunges in a severe undercut that is awe-inspiring.

At all levels of the wall there are clusters of red, orange, and canary yellow tube sponges. At the end of the dive, you should move up on the long shallow gradient above the dropoff; as you move shoreward the coral thins out, and you may find octopus, large hermit crabs, and other odd species moving about on open scoured white limestone.

Wherever you dive along the Bloody Bay Wall, your peregrinations will be rewarded. Weather permitting, you'll dive a truly world-class reef.

Typical depth range	:	15 feet (5 meters) to unlimited (wall)
Typical current conditions	:	None to moderate
Expertise required	:	Novice or better
Access	:	Shore and boat

Of the three islands that comprise the Caymans, Cayman Brac is the only one to feature high ground. The Bluff, which rises as much as 180 feet (55 meters) at its eastern end, is a massive rampart rising high above the sea. This feature gives Cayman Brac a vivid contrast with the low-lying terrain of Grand Cayman and Little Cayman.

Underwater, the terrain of Cayman Brac is a miniature of Grand Cayman. It has its virgin, steep wall on the south coast, near the point

Cayman Brac and Little Cayman are the least dived of the Cayman Islands. Little Cayman has no on-shore dive facilities, while Cayman Brac boasts two. Octopusses are common in cubbyholes of the shallow spur and groove reefs on the north shore of "The Brac."

where the Bluff begins to rise. It has shallow elkhorn coral gardens off its southwest shore that are similar to those off South Sound on Grand Cayman. Cayman Brac even has its own Tarpon Alley, similar in location to the Original Tarpon Alley near the southwestern corner of the island. It has its anchor site on the Southern Wall at 100 feet (30 meters). It even has its North Coast Wall, a series of sites similar to those of Grand Cayman's North Wall—deep reef crests lying 45–50 feet (14–15 meters) beneath the surface, shallow sandy valleys perpendicular to shore, a wall that plunges to unimaginable depths.

There is nothing strange in this similarity. The same tectonic forces, weather, and ocean currents that shaped the reefs of Grand Cayman worked on the smaller cousins to the northeast as well. If anything, Cayman Brac's dive sites are testament to the sea's remarkable consistency in a given area.

Fish life, including queen angelfish, is just as common on Cayman Brac as along the shores of its sister islands. The unfathomable expanse of massive elkhorn gardens along the south shore, in particular, attract tropicals the way an oak forest attracts squirrels.

3

Marine Life of the Cayman Islands

The Cayman Islands are geologically the tips of immense sea mounts rising from the floor of the Caribbean. While they may seem remote their waters are integrally of the Caribbean's circulatory system. The larval stages of species from other Caribbean islands mature here, and Caymanian species dispatch pelagic larvae which currents carry to distant reefs elsewhere in the Caribbean.

The Caribbean is an isolated tropical sea. On the east it is bounded by the windward and leeward island chain and the relatively cold waters of the deep Atlantic. On the west, the land bridge between the North American and South American continents separates the Caribbean from the mighty Pacific.

This was not always so. Over eons, this land bridge we call Central America has been raised and lowered several times. Historically, Caribbean species originated from the same stock as marine life in the Pacific or Red Sea, but millions of years of physical separation have led to minor evolutionary divergence.

For example, many species of butterflyfish in the Pacific or Indian Oceans are rather larger than their Caribbean cousins. Indeed, some of these butterflyfish are as large as the angelfish of the Caribbean. For species such as angelfish and butterflyfish, the physical separation of the Caribbean has led to a totally different identity. The Queen angelfish or the four-eyed butterflyfish, trademark species of the Caribbean, are found in no other sea in the world.

At the other end of the evolutionary spectrum are species which are basically the same world-wide; species as diverse as cornetfish, spiny pufferfish, sharks and manta rays vary little around the world, their long-lived pelagic larvae circulate widely, while those of the butterflyfish and angelfish survive only within the limited circulation pattern of the Caribbean itself.

Nature in its complexity is not content to propagate only local or worldwide species. As you get to know the Caribbean fish and corals more closely, you'll discover subtle local variations. One of the most interesting is in the Royal purple gramma *Gramma Loreto*. Basically this dainty two-inch-long fish has a purple anterior and yellow posterior; yet in the

Fish which live their entire life cycles on shallow reefs often exhibit remarkable diversity of size and coloration throughout the Caribbean. Royal grammas, for example, are marked with varying proportions of purple and yellow according to their location. ▶

Southern Caribbean the purple portion dominates while in the Caymans the yellow portion extends well forward on the body.

Even more complex is the situation among the wrasse and parrotfish species. These fish not only have different color patterns and sizes for juveniles, mature males and females, and breeding supermales; they inter-breed and form hybrids whose color varies across the spectrum.

The novice diver may thus be forgiven for initially finding tropical reef fish a colorful, incomprehensible tapestry of shapes, sizes and colors. Yet, as experience accumulates, the diver discovers the dominant patterns which indeed distinguish the families and species from each other.

There are the sand dwellers such as the jawfish, lizardfish, sand tilefish and many small blennies, who live out their lives on what looks to us like empty white sand sea floor. The sand tilefish even goes about carrying pieces of coral rubble in its teeth, and carefully builds burrows of this rubble. When approached, this foot-long white fish dives into its burrow and disappears.

There are the free swimmers such as horse-eyed jacks, bar jacks, eagle rays, sharks and bonnet mouths. These species travel in open water above the reef, travelling much larger ranges than the local species in their quest for food. At a more local level, close to and even among the coral heads, wander such species as the creole wrasse. Schools of these busy

Black coral, a precious coral used in making jewelry, has become rare within the depth range of sport divers throughout much of the Caribbean. In the Caymans, large trees of black coral can be found as shallow as 20 feet (six meters).

One of the most daring of Caribbean reef fish, the Spanish hogfish is missing from heavily-fished shallow reefs throughout the region. In the waters of Cayman, however, they thrive.

blue fish bustle about on the reef in highly social gatherings which never seem to cease their eternal wanderings.

By far the largest variety of commonly seen species inhabit the immediate vicinity of the coral reef structure itself. For many of these reef animals the corals offer food as well as shelter from predators. In the endless hunt that is life on a coral reef; the nooks and crannies of the complex reef structure are the very key to survival. Here amid and on the corals we find the tiny gobies, the grammas, the butterflyfish and angelfish, the varied wrasses and parrotfish, the schooling snappers and grunts, and the solitary groupers.

As you gain more diving experience, however, you'll begin to discover the many shy species who do not boldly march across the reef. These include the shy drums and croakers, lobsters and crabs, flounders, porgies, sweepers, bigeyes, triggerfish, burrfish and a host of others. It is important to realize that these fish are *not* "rare". They are plentiful, as common as butterflyfish or angelfish, but their behavior pattern is more sheltered, and cautious. For this reason you may not encounter these species until you have become more experienced, and less intrusive. Looking at it from the perspective of a small fish, we divers must seem terrifying monsters indeed. It's no wonder that many species dive into their coral shelters at our noisome approach.

4

Safety

When divers travel to a major destination such as the Caymans, they should keep the principles of good diving practice firmly in mind.

Every famous reef area in the world has its own reef profile, its own cautions. Some destinations such as Palau or the Coral Sea have reeftops which grow to within a few feet of the surface; these offer very easy, safe diving. It's hard to go wrong when you can follow the reef wall right back to the boarding ladder.

A second major element of good diving practice on Caymans' reefs is careful adherence to the decompression tables during the reef portion of the dive. That requires that you remain aware of your depth and of the passage of time. It is a careless diver indeed who becomes so involved in coral and fish that the safe margin of air supply and decompression time dwindles. I have now spent the past fifteen years diving over 300 times per year; believe me, there is nothing that can't wait until the next dive, nothing worth risking the consequences of carelessness. There is always tomorrow to come back and see more.

The third major element of good diving practice is the return to the boat. For one thing, you should plan an overall route for your dive. In a place such as Cayman such routes suggest themselves. At Hole in the Wall or Tarpon alley you would descend from the boat and follow the prominent coral gullies to the main dropoff; the same gullies would later lead you right back to the boat.

If the boat were directly above the wall, my suggestion is to use the wall itself to map your route. Use your first 700 pounds of air to swim south along the wall, your second 700 pounds of air returning north at a

The walls of the island are an open invitation to engage in deep decompression dives. Unless you have been trained in deep diving techniques, resist the temptation to go beyond the no-decompression limits. Hanging off on a line, waiting for the nitrogen to subside before surfacing, is not a practice for novices. ▶

shallower depth. Now you are back beneath the boat with 700 pounds of air or more. On Caymans deep-topped walls no-decompression limits by now are requiring that you ascend to shallower water.

Most experienced divers spend a few minutes hanging at a ten-foot depth whether or not the no-decompression tables require it. A pause to let your system reacclimate can do no harm and much good.

Don't be a "hot dog", or show-off. Going too deep, staying too long, ascending too quickly are invitations to trouble.

Grand Cayman does have a recompression chamber donated some years ago by the local British Sub-Aqua Club after a fund-raising drive. It is a three-man chamber with an In/Out lock. It is located in a separate building at Dr. Poulson's clinic, near the airport.

It is to be sincerely hoped that none of you will ever use that information. Believe me, you can dive the great walls of the Caymans for years and never need to see a recompression chamber. In almost every case where the chamber has been needed, someone bent the rules of elapsed time at depth. If you plan your dive, then monitor your progress regularly against your plan, your margin of safety will be adequate at all times.

A large green moray eel is another occupant of the reef. At night, eels and other nocturnal creatures abound.

DAN. The Divers Alert Network (DAN), a membership association of individuals and organizations sharing a common interest in diving safety operates a **24 hour national hotline, (919) 684-8111** (collect calls are accepted in an emergency). DAN does not directly provide medical care, however they do provide advice on early treatment, evacuation and hyperbaric treatment of diving related injuries. Additionally, DAN provides diving safety information to members to help prevent accidents. Membership is $10 a year, offering: the DAN *Underwater Diving Accident Manual*, describing symptoms and first aid for the major diving related injuries, emergency room physician guidelines for drugs and i.v. fluids; a membership card listing diving related symptoms on one side and DAN's emergency and non emergency phone numbers on the other; 1 tank decal and 3 small equipment decals with DAN's logo and emergency number; and a newsletter, "Alert Diver" describes diving medicine and safety information in layman's language with articles for professionals, case histories, and medical questions related to diving. Special memberships for dive stores, dive clubs, and corporations are also available. The DAN Manual can be purchased for $4 from the Administrative Coordinator, National Diving Alert Network, Duke University Medical Center, Box 3823, Durham, NC 27710.

DAN divides the U.S. into 7 regions, each coordinated by a specialist in diving medicine who has access to the skilled hyperbaric chambers in his region. Non emergency or information calls are connected to the DAN office and information number, (919) 684-2948. This number can be dialed direct, Monday-Friday between 9 a.m. and 5 p.m. Eastern Standard time. Divers should not call DAN for general information on chamber locations. Chamber status changes frequently making this kind of information dangerous if obsolete at the time of an emergency. Instead, divers should contact DAN as soon as a diving emergency is suspected. All divers should have comprehensive medical insurance and check to make sure that hyperbaric treatment and air ambulance services are covered internationally.

Diving is a safe sport and there are very few accidents compared to the number of divers and number of dives made each year. But when the infrequent injury does occur, DAN is ready to help. DAN, originally 100% federally funded, is now largely supported by the diving public. Membership in DAN or purchase of DAN manuals or decals provides divers with useful safety information and provides DAN with necessary operating funds. Donations to DAN are tax deductible as DAN is a legal non-profit public service organization.

APPENDIX 1:

Cayman Dive Services

LIVE-ABOARD DIVE CRUISER

M.V. Cayman Diver II
c/o See and Sea Travel
680 Beach Street, Suite 340/Wharfside
San Francisco, CA 94109; 9-2007

HOTEL BASED DAY BOATS

Bertmar Aqua Sports, Ltd.
P.O. Box 637
Grand Cayman, B.W.I.; 92514/92066

Brac Aquatics
Brac Reef Hotel
Cayman Brac, B.W.I.; 8-7323

Bob Soto's Diving Ltd.
Holiday Inn
P.O. Box 1801
Grand Cayman, B.W.I.; 7-7444

Buccaneer's Inn
P.O. Box 68
Cayman Brac, B.W.I.; 8-7257

Cayman Diving Lodge
P.O. Box 11
Grand Gayman, B.W.I.; 7-7555

Cayman Kai
P.O. Box 1112
Grand Cayman, B.W.I.; 7-9556

Kingston Bight Lodge
Little Cayman, B.W.I.; 8-3244

Spanish Cove Diving Resort
P.O. Box 800
Grand Cayman, B.W.I.; 9-3763

Sunset Divers Ltd.
P.O. Box 479
Grand Cayman, B.W.I.; 9-5966

Surfside Water Sports
Galleon Beach Hotel
P.O. Box 891
Grand Cayman, B.W.I.; 7-4224

The Tortuga Club, Ltd.
P.O. Box 496
Grand Cayman, B.W.I.; 7-7551

INDEPENDENT DAY BOATS

Cayman Undersea Adventures
P.O. Box 151
Grand Cayman, B.W.I.

Clint Ebanks' Scuba Cayman
P.O. Box 746
Grand Cayman, B.W.I.; 9-3873

Fisheye Photographic Services
P.O. Box 637
Grand Cayman, B.W.I.; 7-4209

F.L.A.G.
P.O. Box 446-G
Grand Cayman, B.W.I.; 9-2606

Kingston Bight Lodge
Little Cayman, B.W.I.; 8-3244

Peter Milburn's Dive Cayman Ltd.
Box 596
Grand Cayman, B.W.I.; 7-4341

Seasports
P.O. Box 1516
Grand Cayman, B.W.I.; 9-3965

Ron Kipp's Bob Soto's Diving Ltd.
P.O. Box 1801
Grand Cayman, B.W.I.; 9-2022

Quabbin Dives Ltd.
P.O. Box 157
Grand Cayman, B.W.I.; 9-5597

HOTEL ACCOMMODATIONS

Grand Cayman

Ambassadors Inn	9-5515
Beach Club Colony	9-2023
Caribbean Club	7-4099
Cayman Diving Lodge	7-7555
Cayman Islander	9-5528
Cayman Kai Resort	7-9556
Coral Caymanian	9-4054
Holiday Inn	7-4444
Le Club Cayman	7-4000
London House	7-4060
Paradise Manor	9-5677
Royal Palms Hotel	9-2636
Seaview Hotel	9-4990
South Cove	9-2514
Spanish Bay Villas	9-3272
Spanish Cove	9-3765
Sunset House	9-5966
Tortuga Club	7-7551

Cayman Brac

Brac Reef Hotel	8-7323
Buccaneer's Inn	8-7257
Tiara Beach Hotel	8-7313

Little Cayman

Kingston Bight Lodge	8-3244
Pirates Point	7-4324
Southern Cross Club	None

Appendix 2: Further Reading

Hirst, George, S.S. *Notes on the History of the Cayman Islands.* Kingston, Jamaica: P. A. Benjamin Manufacturing Company, 1970.

Humann, P., and Picairn, F. *Cayman Underwater Paradise.* Bryn Athyn, Pa: Reef Dwellers Press, 1979.

Murphy, G. "Cayman: The Mt. Everest of Diving Adventure." *Skin Diver,* February (1984), pp. 65-101.

Roberts, Harry H. *Field Guidebook to the Reefs and Geology of Grand Cayman Islands, B.W.I.* Miami Beach, Fla.: Atlantic Reef Committee, 1977.

Smith, Roger, C. "Archaeology of the Cayman Islands." *Archaeology,* Vol. 36, No. 5, September/October (1983), pp. 16-24.

Williams, N. *A History of the Cayman Islands.* Grand Cayman: Government of the Cayman Islands, 1970.

Index